T0279514

Advance Praise for *Skilletheads*

"Obviously Ashley has a problem, and thankfully she's kind enough to share it with us! This book really is for everyone. The home cook to the Michelin chef. The newbie collector to the seasoned pro (ha!). A must-have for the cast iron user in your life!" **Kevin Fogarty, Cast Iron Kev**

"*Skilletheads* is likely the most comprehensive, thoroughly researched guide on vintage and modern cast-iron cookware. For anyone interested in exploring the storied history of cast-iron cookware in the United States, this is a great place to start." **Peter Huntley, Stargazer Cast Iron**

"What a great resource! With insights into vintage cast iron and firsthand accounts from modern makers leading the cast iron resurgence, *Skilletheads* is the definitive guide to cast-iron cookware." **Michael Griffin, FINEX**

"This book is a fantastic cross-section of the diverse collector community. It's a perfect justification that new collectors can show to their spouses to say, 'Look! I could be as bad as THESE guys!' Absolutely recommend for anyone who loves cast iron." **Ken Margraff, Cast Iron Savannah**

"If you want to learn more about cast-iron cookware, this book might be your bible. Ashley traces the history of the cast-iron cookware industry, shows us how to restore and care for our cast iron, and leaves us with some amazing recipes so that we can put our favorite pans to use. This book is a celebration of what we Skilletheads love about our tools." **Isaac Morton, Smithey Ironware**

"This book will either help you understand the addiction or start a new one! This is a great place to reference any questions you might have about cast-iron cookware." **Matthew Bright, Orphaned Iron**

SKILLETHEADS

Photo by Courtney Wahl.

A Guide to
Collecting and Restoring
Cast-Iron Cookware

SKILLETHEADS

Ashley L. Jones

RED ⚡ LIGHTNING BOOKS

Photo by Courtney Wahl.

This book is a publication of

RED ⚡ LIGHTNING BOOKS

1320 East 10th Street
Bloomington, Indiana 47405 USA

redlightningbooks.com

This book is printed on acid-free paper.

Manufactured in Canada

First printing 2023

Library of Congress
Cataloging-in-Publication Data

Names: Jones, Ashley L., author.
Title: Skilletheads : a guide to collecting and
    restoring cast-iron cookware / Ashley L. Jones.
Description: Bloomington, Indiana : Red
    Lightning Books, [2023] | Includes
    bibliographical references and index.
Identifiers: LCCN 2022048261 (print) | LCCN
    2022048262 (ebook) | ISBN 9781684352029
    (hardback) | ISBN 9781684352036 (ebook)
Subjects: LCSH: Skillet cooking. | Cast-iron. |
    LCGFT: Cookbooks.
Classification: LCC TX840.S55 J66 2023
    (print) | LCC TX840.S55 (ebook) |
    DDC 641.7/7—dc23/eng/20221122
LC record available at https://
    lccn.loc.gov/2022048261
LC ebook record available at https://
    lccn.loc.gov/2022048262

Contents

My family, February 2022. *Photo by Courtney Wahl.*

Acknowledgments

Skilletheads is what I call a "contribution-based book," one that wouldn't exist without the information, images, and recipes contributed by cast-iron pros, from manufacturers to restorers and foodies. To all of those Skilletheads who spent so much time contributing to this book, thank you.

To Sam Rosolina, PhD, thank you for taking the time to research chemicals and restoration techniques. You gave us answers to questions that restorers have been asking for a long time. I think that means you are now officially a Skillethead.

To Kelli Heil, illustrator extraordinaire, thank you for helping me capture the beauty of cast iron.

Thanks to the folks at IU Press and Red Lightning Books for the chance to share more about cast-iron cookware. I can't believe my good fortune in being able to work with such a great team.

Thank you to my Page 5 writers group through Word Weavers International. I know it wasn't fun to read a bunch of facts and restoration methods out of context every month, but your feedback kept me on track. I think you'll enjoy the end result.

Finally, but most importantly, I must thank my ever-patient husband, Robby, and little boy, Gordon. I wouldn't be writing anything if it weren't for your efforts and encouragement.

SKILLETHEADS

Photo by Courtney Wahl.

Introduction

Skillethead (ski-let hed)

1. A person who spends an inordinate amount of time hunting for rare and quality cast-iron pieces, restoring cast iron, selling restored pieces, and/or collecting cast iron.

2. Someone who participates regularly in online forums and groups devoted to cast iron research, collection, and/or restoration, or someone who has started such a group.

3. Someone who has founded a cast-iron manufacturing or seasoning company.

4. Someone who thinks about cast iron way too much.

Before *Skilletheads*, there was *Modern Cast Iron: The Complete Guide to Selecting, Seasoning, Cooking, and More*. As the title implies, this was written as a complete guide to cast-iron cookware. But there were two big topics it only briefly addressed: collection and restoration. And that is what I've set out to explore in this book.

To cover these topics, I've had to dig into the history of cast-iron manufacturing, research restoration techniques, and interview dozens of modern manufacturers, restorers, and foodies, and one very helpful chemist. Whether you're interested in finding the perfect skillet for your kitchen or you'd like to start restoring cast iron as a hobby or side job, then this book is for you. In these pages, you'll find side-by-side comparisons of modern companies,

For a full bibliography and additional resources, visit my website at ashleyljones.com.

step-by-step restoration guides, and helpful collection and restoration tips from the pros. And no cast iron book is complete without recipes, so you'll find plenty of those as well, all contributed by the people who know cast iron the best.

But I do have to warn you: if you're not already interested in collecting and restoring cast iron, you will be after reading this book. So gas up the car and clear a few Saturdays on the calendar, because you're about to have some fun with cast-iron cookware!

You might be a Skillethead if . . .

1. You spent your vacation traveling out of state to purchase a bunch of cast iron from an old collector you met online.

2. You look for creative ways to use your cast iron collection, such as hanging them in the garden as wind chimes.

3. You've had heated discussions with novices over the use of hand drills (which are a big NO-NO).

4. You have a strong opinion about "textured versus smooth surface" and you're ready to share it with anyone who will listen.

5. You spend all your extra money on "rescuing" rust buckets.

6. You have a lot of pieces in your cast iron collection . . . and they all have names.

7. You're a member of at least one online group or forum devoted to cast iron cooking, collection, or restoration (or you've started one yourself).

8. You've been caught diving into at least one dumpster to get a better look at a rusted pan.

9.   You have a cast iron nickname, which you use with pride.

10. Your family thinks you have a problem . . . and there may have been talk of an intervention at one point . . . but they sure do enjoy your cooking and the pans you give them.

Skillethead Supporters

The family members and friends of Skilletheads often don't understand the allure of collecting and restoring rust buckets. At this time, there are no support groups for these individuals. However, Skilletheads tend to be a genial bunch who love to cook, so it's generally believed (among Skilletheads, at least) that the benefits of receiving restored cookware and eating homemade cast iron dinners should offset any frustration or embarrassment from being associated with a Skillethead.

Photo by Courtney Wahl.

Collecting Vintage Cast Iron

"They don't make 'em like they used to." That old adage is certainly true for cast-iron cookware.

Before the 1960s, cast iron was molded, poured, and polished by hand. The tradesmen were often paid by the piece, so they worked quickly but competently, ensuring their pieces would pass inspection. Many of these old pans and ovens bear the mark of the foundrymen who made them, identifying each as a unique work of art.

Vintage pans often prove to have thinner walls, lighter weight, and finer craftsmanship than pans manufactured through automation, which began in the early 1960s. That's because the new machinery was rough on the pan, requiring thicker walls, stronger handle connections, and smaller pour spouts.

If you want a one-of-a-kind, tangible piece of history, then look for a vintage cast-iron pan. I've highlighted some of the major vintage manufacturers in this section. None of these companies exist today, but you can still find their products available for resale.

"The Iron Foundry, Burmeister and Wain," an 1885 painting by Dutch painter Peder Severin Krøyer. Now located in the Statens Museum for Kunst in Copenhagen.

Where Is the Cast Iron?

You may think that cast iron is a southern thing, or found only in New England, or restricted to the plains of the Midwest. When I spoke with Isaac Morton, founder of Smithey Ironware, he made this comment:

"All of our customers believe that cast iron is native to them. And it speaks to the importance of it as being kind of a tool that reaches back to certain memories. Every region claims cast iron for its own."

Cast-iron cookware was introduced in the United States by our founding fathers when they established the first American colonies. Foundries were later established in areas rich in iron ore, mainly in the Northeast and Midwest. But from there, cast-iron pans and Dutch ovens traveled all over the continent, feeding hungry cowboys and land speculators alike. That's why you can still find cast-iron cooking—and cast-iron pans for sale—in virtually every city in America.

Map of Cast Iron Restorers and Manufacturers Listed in this Book

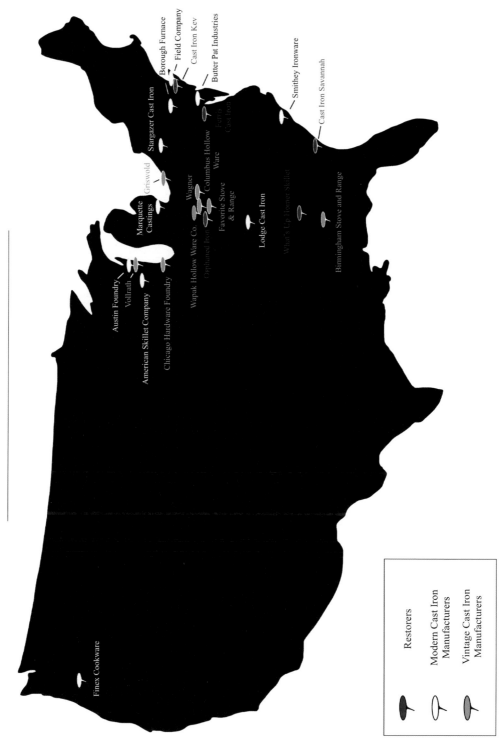

Borough Furnace
Field Company
Cast Iron Key
Butter Pat Industries
Smithey Ironware
Cast Iron Savannah
Stargazer Cast Iron
Ferris Cast Iron
Griswold
Columbus Hollow Ware
Wagner
Marquette Castings
Wapak Hollow Ware Co.
Favorite Stove & Range
Orphaned Iron
Lodge Cast Iron
What's Up Homer Skillet
Birmingham Stove and Range
Austin Foundry
Vollrath
American Skillet Company
Chicago Hardware Foundry
Finex Cookware

Restorers

Modern Cast Iron
Manufacturers

Vintage Cast Iron
Manufacturers

Shopping for Vintage Cast Iron

To truly appreciate your new pan, it should be comfortable to handle—not too heavy or bulky for you to lift. It should also appeal to your aesthetics, for you're sure to use a pan you enjoy more than one you don't. Lastly, you should appreciate the history of your vintage pan, and for that you'll need to research the manufacturer.

If you're interested in restoring and selling cast iron, then you'll need to become a bit of an expert on cast-iron cookware so you'll know which pieces to purchase and at what prices to sell them. (Read more about the restoration process itself in "Restoring Cast Iron.")

There are many resources available, but these are the ones most recommended by the Skilletheads I interviewed:

- ***A Cast Iron Journey* by James P. Anderson** This small book provides an introduction to collecting and restoring cast iron and includes sample photos of pans from major manufacturers.

- ***The Book of Griswold & Wagner* ("the Blue Book") and *The Book of Wagner & Griswold* ("the Red Book") by David G. Smith and Chuck Wafford** These two books are considered the "bibles" of cast iron collection and are on every restorer's bookshelf. Each book contains detailed images and manufacturing years for products from the most famous companies, including Wagner, Griswold, Favorite, Wapak, Lodge, and Vollrath. While these books are excellent for identification purposes, consider their valuations as starting points only. If you're trying to sell restored cast iron, consult online sales pages for current pricing of similar cast-iron pieces.

- **CastIronCollector.com** This website contains a trove of information on cast-iron cookware as well as images to help you identify pieces. To connect with other cast iron fans and to post your own questions, visit their Facebook group at Facebook.com/CastIronCommunityCIC.

WHERE TO SHOP

With the increasing interest in cast-iron cookware, vintage pieces are becoming harder to find. Here are the places Skilletheads frequent along with tips on getting the best deal:

- **Antique/consignment shops** These stores usually have cast iron, but the prices are high.

- **Auctions** Look over the inventory and determine how much you're willing to spend before the auction begins.

- **Estate sales** Look for entire collections of cast iron.

- **Facebook Marketplace** This is a big resource for many Skilletheads.

- **Flea markets** You can still find cast iron here, but as Orphaned Iron says, "You can't be afraid to haggle!"

- **Garage sales / yard sales** You may find low-priced pieces that have been well cared for, but they're few and far between. Cast Iron Kev recommends shopping only at larger combined sales because they're more likely to have cast iron and you'll waste less time and money on driving.

- **Online sales sites like Craigslist and eBay and apps like OfferUp** Look for bulk sales from retired collectors.

- **Swap meets** Connect with local cast iron collectors in your area and with groups online to learn when the next swap meet is scheduled.

- **Thrift stores** These stores don't always have cast iron, but when they do, it's often sold at a low price. Introduce yourself to the owners of the local stores and ask them to notify you when they receive cast iron donations. If you offer to pay a little extra, they may be willing to make the effort.

- **You name it** One Skillethead told me he found a piece of cast iron in a snowbank! If you keep your eyes open and let others know you're on the hunt, you may be pleasantly surprised. As What's Up Homer Skillet told me, "Finding a piece where one is least expected is thrilling."

"Nobody drives farther, gets there earlier, or walks faster than me!"

—Cast Iron Kev

WHAT TO LOOK FOR

If you're in the market for vintage cast iron, you can't be afraid of rust. In fact, Skilletheads often look for the dirtiest, rustiest, cruddiest piece of iron they can find because it's cheaper than a pristine pan. They're going to strip and re-season it anyway.

But what's under all that rust? Learn as much as you can about the piece before you buy it. If you're shopping online, the seller should be able to answer basic questions about the pan. If you're shopping in person, though, you get to be the detective.

Your tools:

- A small flashlight

- A pocketknife

- A measuring tape

- A straightedge or ruler

- A small magnet

- Collector's guidebooks (e.g., the Blue and Red Books)

- Smartphone with internet and camera functionality

PRO TIP When purchasing cast iron online, beware of scams. Request additional pictures of the cast iron to be sure it's in their possession. Ask how they plan to ship it and if it will be insured, and check their sales ratings, if available. In general, established restorers (like those referenced in this book) and cast iron sales groups are safer than one-off sales from unknown individuals.

To determine the quality of the pan, look for:

- **Cracks** Use your fingertips to feel all over the cast iron, and follow up with a close visual inspection with your flashlight. Cracked pans cannot be restored or used safely, so avoid these, regardless of brand. (That is, unless you plan to cut them up and make cast-iron spatulas out of them like Cast Iron Kev does!)

- **Chipping** If the pan has a small chip, it may still be usable, but its value will be reduced.

- **Pitting** This could be a result of previous rust and restoration, and it can keep the pan from accepting seasoning.

- **Heat damage** People often use fire to clean cast iron, but the extreme heat can cause thermal shock to the pan, causing it to warp, making it more brittle, and changing the composition of the iron so it refuses to keep a good seasoning.

To determine if a pan has heat damage, first look at the color. If it has a red or pink hue, it was likely in a fire. Then use your straightedge on the bottom of the pan or place the pan on a level table and see if it wobbles or spins. Place the straightedge on the inside of the pan to see if it is warped at an upward angle. Warped pans, wobblers, and spinners cannot be used on a smooth-surface stove top, but they may be used on a gas-top stove, on a grill, or over an open fire. If it is an otherwise good piece of cast iron with a well-known brand, you may be able to restore and sell it, but you will need to disclose the damage and charge a much lower price. Better yet, keep it for your camping needs.

To identify the pan, get your guidebook out and note the following:

- **Material** Sometimes it can be difficult to determine by sight alone if a pan is made of cast iron or aluminum. Use your magnet to test the pan. A magnet will stick to iron but not to aluminum.

- **Measurements** The size of a pan is measured by the diameter of the top rim (not the bottom), while Dutch ovens are usually measured by the quart. Even the weight of the pan is a clue.

- **Shapes** Pay close attention to the shape of the handle and the pour spouts, if there are any. Each company and each brand will have its own design.

- **Markings** Use your flashlight to look for any numbers, letters, or symbols on the top of the handle, under the handle, and on the bottom of the pan. Also, check for a heat ring around the edge of the bottom of the pan. If present, is the heat ring in one piece, or does it have notches?

All of these clues will help you identify the pan using your guidebook. If the rust is too thick, you may be able to remove a bit using a pocketknife, but check with the seller first. If no markings are visible, it may be because there were none to begin with or because they've been removed through wear or previous restorations. Or the rust may simply cover the markings until you restore the piece.

If you're still unsure about the pan, Cast Iron Savannah recommends taking a picture and posting it in an online forum like the Cast Iron Community Facebook group. Someone within the group may be able to tell you more about the piece before you purchase it.

Does this sound like a lot? Don't worry—it becomes second nature. As Cast Iron Kev said, "Once you've been doing this long enough, you can identify 80 percent of pieces just walking by! Shape of the handle, heat ring, weight, pour spouts—once you're in the groove, you can get a good idea of what it is even covered in crust!"

"Start small and be realistic. Find a foundry that 'speaks' to you based on if you like the way it looks or feels. Then start pickin'!"

—Cast Iron Kev

PRO TIP What's Up Homer Skillet recommends buying larger pieces of cast iron if you plan to resell them. He says waffle irons and kettles are really neat but don't sell as well and are more difficult to restore.

WHAT TO AVOID

Not all vintage cast iron is worth collecting or valuable enough to resell. While foreign pieces are replete in the marketplace, they're not considered to have the same level of quality craftsmanship as American-made cast iron, so most Skilletheads avoid them.

If you plan to restore and resell cast iron, you may want to avoid pieces made after the introduction of automation, which is between the 1950s and 1960s,

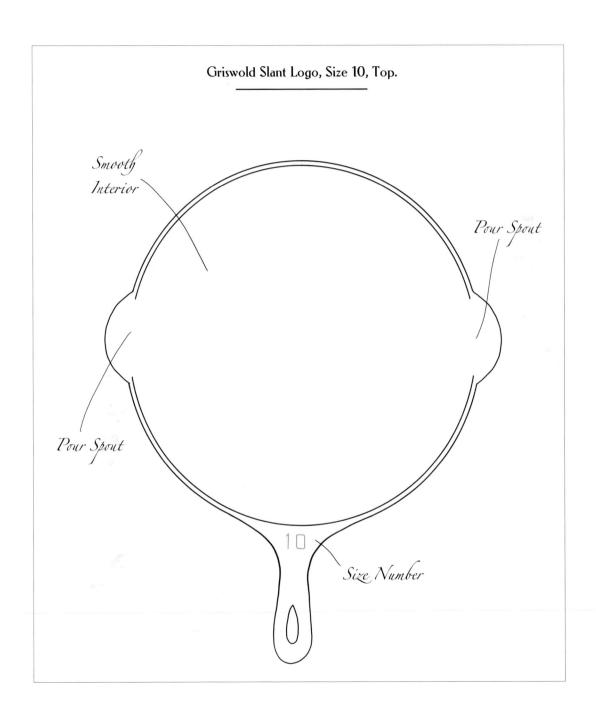

Griswold Slant Logo, Size 10, Top.

Smooth Interior

Pour Spout

Pour Spout

10

Size Number

Griswold Slant Logo, Size 10, Bottom.

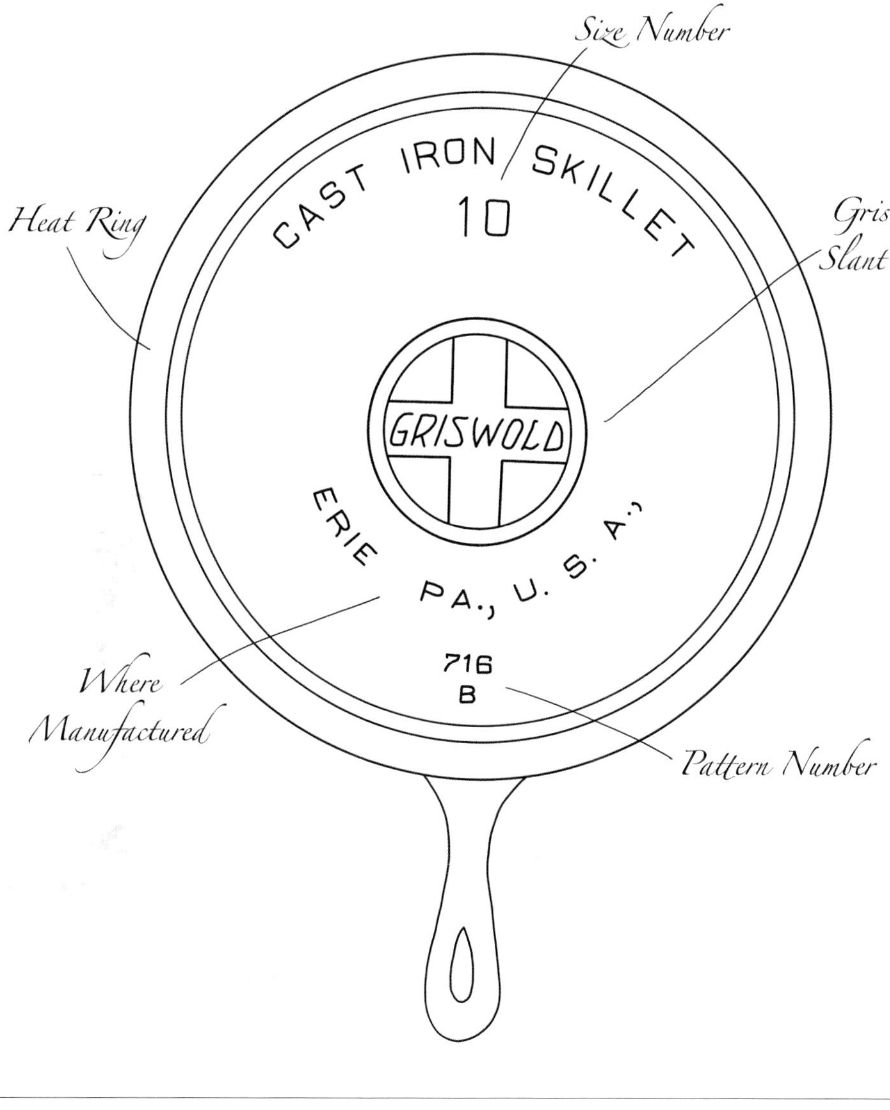

Size Number

Heat Ring

CAST IRON SKILLET
10

Griswold
Slant Logo

GRISWOLD

ERIE PA., U. S. A.,

716
B

Where
Manufactured

Pattern Number

depending on the company. Do your research and focus on the handcrafted earlier pieces to earn a premium.

Vintage Terminology

When researching vintage cast iron, you'll come across a lot of new terms. Here are a few descriptions to help you get started.

AUTOMATION

Manufacturing cast iron was always a very manual process. Foundrymen created sand molds one at a time, which they filled from a ladle of molten iron. As early as the 1930s, companies looked for ways to automate their processes to increase efficiency.

In 1965, Lodge Cast Iron became the first manufacturer to use the automatic DISAMATIC machine on US soil.[1]

The Match Plate system also became available in the 1960s and was favored by manufacturers for shorter series runs. Both the DISAMATIC and the Match Plate system are used widely today.[2]

Since the manufacturers introduced automation at different times and with different processes, each has its own pre- and post-automation dates. If you're collecting or restoring cast iron, you may want to look for pans produced prior to automation, which are usually more favored and more valuable than those produced with an automated process.

FRAUDS

It's sad but true: that pan you're looking at may be a fraud. Domestic and foreign companies alike have copied existing pans and other cast-iron products and passed

"The well-known brand names are always smart buys if the prices are right."

—Cast Iron Savannah

One of the original designs for the DISAMATIC machine. *Courtesy DISA.*

4 star revolution in moulding technique

them off as the real deal. Fortunately, cast iron is a quality-based product, and counterfeits are usually easy to spot.

Check your guidebook to determine how big the pan should be, how much it should weigh, and what exactly the pan should look like, from the handle to the markings and logo. Then look for anything inappropriate with the pan in question, including obvious damage, "rough" casting, or an odd-shaped handle. Even misspellings are common with fraudulent pieces.

GATE MARKS

Gates refers to the place where the cast-iron pan is pinched off from the mold. Early casting used a bottom gate system that created a raised slash or scar on the bottom of the pan. In the mid- to late nineteenth century, side gating was introduced, which moved the gate to the side of the pan, where any remaining marking could be rubbed off.

Bottom gate marks denote older pans from the 1800s. These typically do not have the manufacturer's name on them. It should be noted that Wagner and

Griswold never used bottom gating, so any pan bearing their logos with a bottom gate would be deemed a recast (or fraud).[3]

GHOSTS

Cast iron patterns have never been cheap or easy to make. When a company wanted to remove its logo to sell its products in stores at a lower price or to replace it with a new one, it would often fill in the logo on the original pattern. Over time, the fill would wear away, and the pattern would display both the new and old logos. The pans created by these patterns would then carry both the new logo and a slight "ghost" impression of the old logo.

Another scenario is that one company would use another company's pan as a pattern. This would result in the original company's logo appearing faintly next to the new company's logo.[4]

Pans with ghost logos are not necessarily more valuable than other pans, but they are appreciated as novelties.

HEAT RINGS

A heat ring is a raised ring around the bottom of a skillet, also known as a smoke ring. Manufacturers claimed this was supposed to lift the pan slightly off the eye of a wood stove or—if the eye was removed—out of the flame itself.

Interestingly, that wasn't the real purpose of heat rings, which were originally called machining rings. Making a large cast-iron surface completely flat is quite difficult, so manufacturers added the ring, which was much easier to machine flat. However, some companies removed the heat ring over time or used it only for certain brands.[5]

If you have a solid-surface cooktop, you may prefer a pan without a heat ring to ensure continuous contact with the surface and faster heating time.

HOLLOW WARE

The industry term for cast-iron cookware was *hollow ware*, differentiating it from other cast-iron products, such as cannonballs and flat irons. A set of cast-iron pans and pots was deemed so essential as to even be called furniture at one point in time.

INCISED

If a pattern had a raised symbol or character, that image would be incised, or indented into the cast-iron pan. Most logos and markings were incised.

LOGOS

When companies started adding their logos to their pans, it certainly made it easier to identify the manufacturer. However, these logos changed over time, and their various manifestations help to identify the year the pan was manufactured, the brand name, where it was manufactured, and even where it was sold.

Terms such as *ARC*, *block*, and *slant* describe the way the logo appears on the pan, whether in an arc shape, written in block letters or with a slant font, and so on.

Sandshifting on a Favorite #12 pan.
Courtesy Orphaned Iron.

MINIS

Too small for cooking, mini pans were used for advertising purposes by cast-iron manufacturers. These are not to be confused with cast-iron toy sets, however, which were packaged and sold for children to play with.

NOTCHES

You may have come across the terms *no-notch* and *three-notch* skillets. These terms refer to the number of notches, or spaces, within the raised heat ring and help to identify the manufacturer and the year of manufacture.

MULL & REED.

Spider.

No. 58,460.

Patented Oct. 2, 1866.

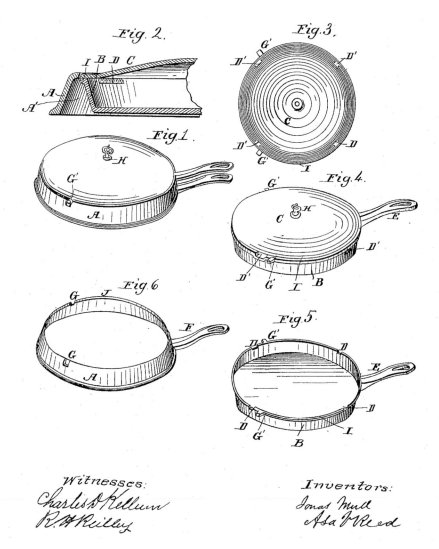

Fig. 2.

Fig. 3.

Fig. 1.

Fig. 4.

Fig. 6.

Fig. 5.

Witnesses:
Charles D. Kellum
R. H. Reilley

Inventors:
Jonas Mull
Asa F. Reed

Illustration for Patent US58460, Spider.

SANDSHIFTS

During the casting process, the sand mold would sometimes shift slightly, causing a disturbance in the iron. These sandshifts (or sags) can make a vintage pan more desirable.

SIZE NUMBER

Cast iron was initially used exclusively in a hearth or over an open fire. But when wood stoves became popular, their manufacturers began to create cast iron in dimensions that matched the size of the stove's eyes, or burners. A wood stove with a Number 8 size eye would require a Number 8 (also written as #8 or No. 8) cast-iron pan.

Since each manufacturer determined its own sizing, a Number 8 with one company could have been a slightly different size than the Number 8 of a competing manufacturer.

SPIDERS

Back when dinners were cooked on a hearth, people often used a three-legged, long-handled frying pan. Due to their long, thin legs and wide, black bodies, these pans were called spiders. When wood stoves entered the market, pans were designed without legs, and the handles were shortened. Still, the name persisted. If you see a cast-iron spider for sale or a recipe for Spider Cornbread, rest assured it has nothing to do with arachnids and everything to do with cast-iron goodness.[6]

Manufacturers

BIRMINGHAM STOVE & RANGE CO

Manufactured in Birmingham, Alabama
Founded by Sam D. Jones
Period of hollow ware production 1902–1993
Brand names Red Mountain, Century, Pioneer, Lady Bess

In 1898, the Jones family acquired controlling interest in Atlanta Stove Works (ASW), which produced wood and coal stoves out of Atlanta, Georgia. They then built a foundry in North Birmingham for the purpose of producing hollow ware for the Atlanta Stove Company. As for foundrymen, the company leased eighty convicts from the State of Alabama. Originally named Alabama Manufacturing Company, the business was renamed Birmingham Stove & Range Co (BSR) in 1909 after acquiring a number of foundry patterns.[7]

In the 1950s, along with other cast-iron manufacturers of the day, BSR began using an automated molding process, followed by high-volume DISAMATIC automated molding machines in 1966.

Although the Atlanta foundry was closed in 1957, BSR continued to manufacture products out of their Birmingham foundry under the ASW and BSR names.

The BSR foundry closed in 1991, and the company sold their molding machines to Robinson Iron, which continued to produce cookware under the BSR name; Lodge Cast Iron handled the distribution. Lodge also produced the popular Sportsman Grill on behalf of BSR. These arrangements were short-lived, however. BSR filed bankruptcy in 1993 and turned over their patterns to Lodge to satisfy their debt.

BSR is credited with the introduction of the divided corn bread skillet in 1967. Although Lodge still holds the BSR patent, their modern version of this pan has two loop handles instead of a single long handle. So if you want a traditional divided corn bread skillet, you'll have to go vintage.

FUN FACT
Ferris's favorite vintage brand is BSR.

BSR #10 Century Series Cast Iron Skillet

Chicago Hardware Foundry Diamond Logo Skillet

CHICAGO HARDWARE FOUNDRY CO.

Manufactured in North Chicago, Illinois
Founded by John Sherwin, E. P. Sedgwick
Period of hollow ware production 1900–1963
Brand names Favorite, Sani-Ware, Ni-Resist

In 1900, two employees of Chicago Hardware Manufacturing Co., John Sherwin and E. P. Sedgwick, had some creative differences with their employer. Still, they were allowed to lease some factory space to make their own cast-iron products. Their business was so successful, they eventually built their own plant across the street from Chicago Hardware Manufacturing Co. In an apparent effort to create confusion, they named their company Chicago Hardware Foundry Co. (CHF). The new company produced cast-iron skillets, among other products, and in 1934 they acquired patterns and tooling from the defunct Favorite Stove & Range Co.

A labor strike at CHF resulted in a riot in 1938. Though no one was seriously hurt, it marked the beginning of a period of strikes and claims of employee abuse that continued for years. At one point, CHF imported fifty Puerto Ricans to work for $5 a week. The Made in Chicago Museum notes these workers were "housed in company-owned railroad cars and forced to buy their food and work clothes at a company store."[8] In addition to the personal rights violations these employees suffered, the company stores served to depress the local economy. When foundry employees were paid only in store credit, they had no money to spend elsewhere, causing businesses in the area to collapse.

It is unclear when CHF stopped producing hollow ware, but we do know the company shifted its focus to industrial furniture in 1963 and was put under the control of a New York holding company in 1969. The surviving equipment was transported to another foundry in Racine, Wisconsin. The Made in Chicago Museum states, "As best we can tell, some bastardized version of the Chicago Hardware Foundry Co. continued to exist as late as 1988, with production in that same Racine foundry."[9]

In 1988, the company relocated to Grayslake, Illinois, and lost a lawsuit brought by the state of Wisconsin for releasing toxic metals. That same year, a chemical fire destroyed the foundry's former North Chicago plant, bringing a violent closure to the company.

COLUMBUS HOLLOW WARE CO.

Manufactured in Columbus, Ohio
Founded by Jesse F. Hatcher, E. B. Hatcher
Period of hollow ware production 1882–1902
Brand name The Favorite

The Columbus Hollow Ware Co. was founded in 1802 and located in the old foundry of the Harker Manufacturing Company. It seems the company ran into financial difficulties in the mid-1880s, most likely as a result of having to compete with the cheap labor at the local Ohio State Penitentiary. That prison housed its own foundry and was able to sell its products at lower prices than private companies. In a case of "if you can't beat 'em, join 'em," Columbus Hollow Ware contracted with the penitentiary to produce its line of cookware called "The Favorite." This brand is easy to spot, as "THE FAVORITE" is incised in all caps in the twelve o'clock position on the back of each pan.

Today, we may be concerned by the idea of forced labor or may wonder at the working conditions to which the prisoners were subjected. However, the concerns of the day were closer to home. This cheap labor put more than one private company out of business, adding to the already high level of unemployment. It seemed to society a great injustice that good, hardworking people should be without an income while prisoners had jobs. One response was a proposal to label these products as "Prison Made," much as we label products "Made in China" today, but that plan never came to fruition.[10]

Although these pans have a rather dark heritage, that doesn't seem to hurt their value. Instead, their unusual history, combined with a short production period, makes them highly sought after by collectors. These pans, which were made with the cheapest labor and sold at a budget price, now fetch top dollar.

Just be sure not to confuse The Favorite brand with the Favorite brand from Chicago Hardware Foundry Co. or Favorite Piqua Ware from Favorite Stove & Range Co.

FUN FACT
Orphaned Iron's favorite vintage pan is The Favorite. He states they are "few and far between, but a truly interesting history behind them."

The Favorite by Columbus Hollow Ware

#2 Favorite Piqua Ware. *Courtesy Skillethead John.*

FAVORITE STOVE & RANGE CO.

Manufactured in Piqua, Ohio
Founded by William King Boal
Period of hollow ware production 1889–1935
Brand names Favorite Piqua Ware, Miami (economy brand),
Puritan (for Sears Roebuck)

In 1848, the W. C. Davis Company was founded in Cincinnati, Ohio. Over time, the company was renamed Great Western Stove Works, then the Favorite Stove Works Company. In 1888, William K. Boal relocated the foundry to Piqua, Ohio, and resumed operations the next year under the name Favorite Stove & Range Co. (FSR).

Boal passed away in 1916, and his son, William S. Boal, succeeded him. He quickly expanded FSR's production of hollow ware. At one point, FSR's operations were spread over ten acres of land, making it the largest manufacturing company in the county. With nearly six hundred employees, it had such an impact on the city of Piqua that it became known as "The Favorite City."[11]

In 1919, employees participated in a ten-day labor strike. Their demand: a 25 percent increase in wages.

The Great Depression of the 1930s reduced sales at FSR, as it did with other manufacturers. William S. Boal passed away in 1933, and the company liquidated its assets two years later. Patents, trademarks, and tools were sold to Foster Stove Company of Ironton, Ohio, while the patterns and machinery were sold to Chicago Hardware Foundry Co.

FSR was restructured under the name Favorite Manufacturing Company. They produced coal and wood ranges, gas cooking stoves, and even hollow ware but on a much smaller scale, with their molding outsourced to a local foundry. In 1959, Favorite Manufacturing Company finally stopped operations.[12]

Skillethead John's favorite cast-iron piece is the #2 Favorite Piqua Ware pan. He doesn't use this little pan for cooking, though. Valued at $200, this gem is a collector's piece!

GRISWOLD MANUFACTURING CO.

Manufactured in Erie, Pennsylvania
Founded by Matthew Griswold
Period of hollow ware production 1885–1957
Brand names Selden & Griswold, Erie, Griswold's
Erie, Victor (economy brand), Griswold, Iron Mountain
(economy brand, unmarked), Good Health (store
brand), Best Made S.R. & Co. (for Sears Roebuck),
Puritan (for Sears Roebuck), Merit (for Sears Roebuck)

FUN FACT Smithey Ironware's 10" and 12" pans are loosely based on the Erie No. 9. Founder Isaac Morton exclaimed, "It's the icon to me of what a cool piece of cast iron is."

Although most hollow ware was manufactured by stove companies, Griswold was one of the few manufacturers who focused on the cookware. This turned out to be a solid business plan as Griswold ultimately became a household name and a world leader in cast-iron cookware production.

In 1865, Matthew Griswold moved from his family farm in Connecticut to Erie, Pennsylvania. There he started a humble business venture with his cousins, the Selden brothers, producing butt hinges and other hardware. The company expanded its production into hollow ware in the 1870s, and Griswold purchased his cousins' interests in the business in 1884.

A devastating fire occurred in the factory in 1885 (a sad but common event among manufacturing companies). Undeterred, Griswold rebuilt the factory and reorganized the business in 1887 as Griswold Manufacturing Co.

Over a period of years, various Griswold family members retained the position of president and oversaw tremendous growth, both domestic and international. In 1946, Ely Griswold sold the family company to a New York investment group. The company was then purchased by McGraw Edison in 1957, only to be sold to the Wagner Manufacturing Co. division of the Randall Company. Griswold pans manufactured after this date are not considered collector's items.

In 1959, Randall sold the rights of both Griswold and Wagner to Textron, Inc., who continued to manufacture Griswold-trademarked products in the Wagner plant in Sidney, Ohio, until 1969. At that time, General Housewares Corp. acquired the rights to both companies. However, the use of the Griswold trademark was discontinued in 1973.

Since the factory was located in Erie, Pennsylvania, the first logo the company used (from 1880–1907) was simply "ERIE." Other logos and markings were later used, including the now-famous cross in a circle containing the word "GRISWOLD."[13]

Erie Spider Logo Skillet by Griswold

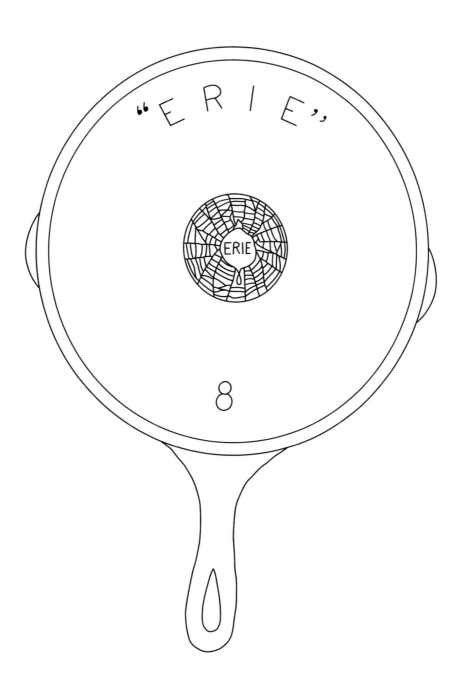

Jacob J. Vollrath Manufacturing Co.'s exhibit in the Palace of Manufacturers at the 1904 World's Fair. *Courtesy Missouri Historical Society.*

VOLLRATH MANUFACTURING CO.

Manufactured in Sheboygan, Wisconsin
Founded by Jacob J. Vollrath
Period of hollow ware production 1884–1960s
Brand name Vollrath

Vollrath Manufacturing Co. was founded as Cast Steel Co. in 1853. Company founder Jacob Vollrath learned iron molding in Germany before emigrating to the United States. Although enamelware was common in Germany at that time, it was hard to come by in America. Seeing an opportunity, Vollrath decided to bring enamelware to the States. There was just one problem: he didn't have the skill set.

Undeterred, Vollrath sent his son, Andrew, to Germany to learn the secrets of enameling. After one failed attempt and a second trip to Germany, Andrew finally became an expert. By 1876, the Vollraths were able to produce excellent quality enamelware, which they made and sold in small batches. The company continued their production of enamelware, formed with either cast iron or stamped steel, into the 1950s.

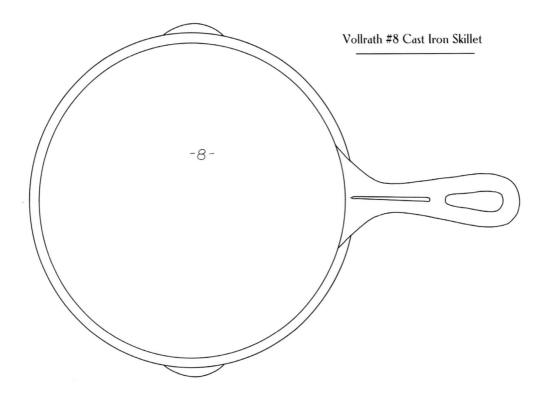

Vollrath #8 Cast Iron Skillet

-8-

The company was renamed Jacob J. Vollrath Manufacturing Co. in 1884 and then simply Vollrath Company in 1912. In 1900, the company devoted itself entirely to the manufacture of cooking utensils, earning top honors for "Excellence in the Production of Colored and Plain, Stamped Steel and Cast Iron Enameled Wares" at the 1904 World's Fair in St. Louis. As with other manufacturing companies, their capacity was diverted to wartime production in the 1940s, but the company persevered. To keep up with innovation, they discontinued enamelware in the 1950s in favor of stainless steel. Finally, cast iron production was discontinued in the 1960s.[14]

Unlike other cast-iron manufacturers, Vollrath continued to transform itself. Today, it is a thriving family business with a history of acquisitions and growing interests in the commercial food service and health-care industries.

Vintage Vollrath cast-iron pans may be found without a logo or incised with "VOLLRATH WARE." Interestingly, the logo and all markings are positioned to the side (with the handle facing the three o'clock position). The handles are recessed with a distinct reinforced ridge down the center, making them easily recognizable.

WAGNER MANUFACTURING CO.

Manufactured in Sidney, Ohio
Founded by Milton M. Wagner and Bernard P. Wagner
Period of hollow ware production 1891–1959
Brand names Wagner, Sidney, Wagner Ware, National (economy brand), Long Life (store brand), Montgomery Ward/Wardway, Ward's Cast Iron, Magnalite (cast aluminum)

When the Wagner brothers founded the Wagner Manufacturing Co. in 1891, they must have had high hopes for their company. Still, they could not have foreseen how extremely popular their brands would become or how in-demand they would be one hundred and thirty years later.

With a look to the future, Wagner added nickel-plated and cast aluminum ware to their catalog in 1892, becoming one of the first manufacturers to do so. In 1897, they purchased Sidney Hollow Ware and then produced the popular Sidney brand of skillets and Dutch ovens. In 1913, Wagner extended their distribution to Europe, a grand feat in the world of cast-iron manufacturing.

"We do not strive to manufacture hollowware as cheaply as possible but as good as it can be made. We cannot afford to put on the market ware that will not sustain our reputation. The name Wagner is cast on the bottom of each piece of ware."

—From an old Wagner advertisement

From 1946 to 1953, the heirs of the original founders began to divest their shares of the company, leading to a rather complicated series of relationships. The Randall Company out of Ohio (an auto parts company) purchased Wagner in 1952. The Wagner division of Randall then purchased a competitor, Griswold Manufacturing, in 1957, only to have Textron, Inc. purchase Randall (including the rights to Wagner and Griswold) in 1959. Collectors mark this as the end of the official production period of collectible Wagner cookware.[15]

Ten years later, Textron sold the Wagner and Griswold lines to General Housewares Corp., which stopped the manufacture of Wagner Ware in 1994. They then sold the rights to Slyman

Wagner #3 pan. *Courtesy Cast Iron Savannah.*

Group in 1996, at which time the Wagner plant fell into receivership. The bank sold the Wagner factory and the Wagner and Griswold trademark rights to American Culinary Corp. in 2014, and they remain there to this day.

FUN FACT Cast Iron Kev's favorite piece of cast iron is a 9" Wagner Chef Skillet while Kent Rollins claims Wardway as one of his favorite vintage brands.

FUN FACT Smithey Ironware's pans have three small holes at the end of the handle. Founder Isaac Morton came up with this idea after seeing an old Wagner pan with three holes across the handle.

WAPAK HOLLOW WARE CO.

Manufactured in Wapakoneta, Ohio
Founded by Milton Bennet, Marion Stephenson, Harry Bennett, Charles Stephenson, S. P. Hick
Period of hollow ware production 1903–1926
Brand names Wapak, Oneta (economy brand)

Records show that Wapak was founded in 1903, but after that little is known. *The Book of Griswold & Wagner* (i.e., "the Blue Book"), records the following: "Even though the Wapak Hollow Ware Co. boasted that it was the largest exclusive cast hollow ware manufactory in the world, and the largest and most important employer of labor in Wapakoneta, there is little recorded history of its existence. In fact, other than the listing in *History of Western Ohio and Auglaize County*, there is almost no published history of this company. County records indicate, however, that the Wapak Hollow Ware Company ended in bankruptcy in 1926."[16]

Wapak is known for the Indian head in their trademark, which was most likely a nod to the rich Native American history of the area. The handles on the Indian Head pans are also unique in shape. While these Indian Head pieces are prized collector's items, for their quality as well as their uniqueness, the other products in Wapak's catalog are interesting for another reason: they appear to be copies of other companies' products. For example, the hinges used in Wapak's waffle makers copy those manufactured by Sidney Hollow Ware and Wagner. As for their pans, pattern markings from Griswold and ERIE ghost markings can often be detected, indicating that Wapak used actual pans from other companies to create their own patterns. Although this was not an uncommon practice of the day, it was certainly frowned upon. And if the patterns of the original pans were patented—as they often were—then it was illegal as well.[17]

For all these reasons, collectors today are quite interested in Wapak's hollow ware.

Wapak Indian Head Medallion close-up. *Courtesy Orphaned Iron.*

Where Did All the Cast Iron Go?

Over time, cast-iron manufacturers had many difficulties to contend with. Those that survived the world wars and the Great Depression then had to battle the introduction of lightweight aluminum, stainless steel, and chemical nonstick cookware, as well as globalization and the availability of lower-priced imports from Asia. Cast-iron manufacturers tried their best to keep up, with many of them introducing their own lines of alternative cookware, but eventually all of them—except Lodge Cast Iron—closed their doors.

The reign of cast iron had come to end.

At least for a while.

FUN FACT Dennis Powell, founder of Butter Pat Industries, acknowledged that their handle gives a "very explicit nod to Wapak Foundry's 'Indian' handle."

Cast Iron Manufacturing Dates

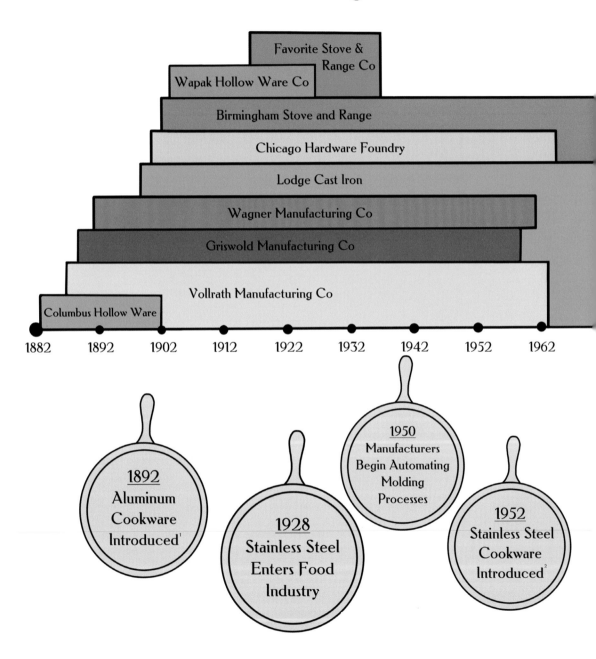

Favorite Stove & Range Co

Wapak Hollow Ware Co

Birmingham Stove and Range

Chicago Hardware Foundry

Lodge Cast Iron

Wagner Manufacturing Co

Griswold Manufacturing Co

Vollrath Manufacturing Co

Columbus Hollow Ware

1882   1892   1902   1912   1922   1932   1942   1952   1962

1892
Aluminum Cookware Introduced[1]

1928
Stainless Steel Enters Food Industry

1950
Manufacturers Begin Automating Molding Processes

1952
Stainless Steel Cookware Introduced[2]

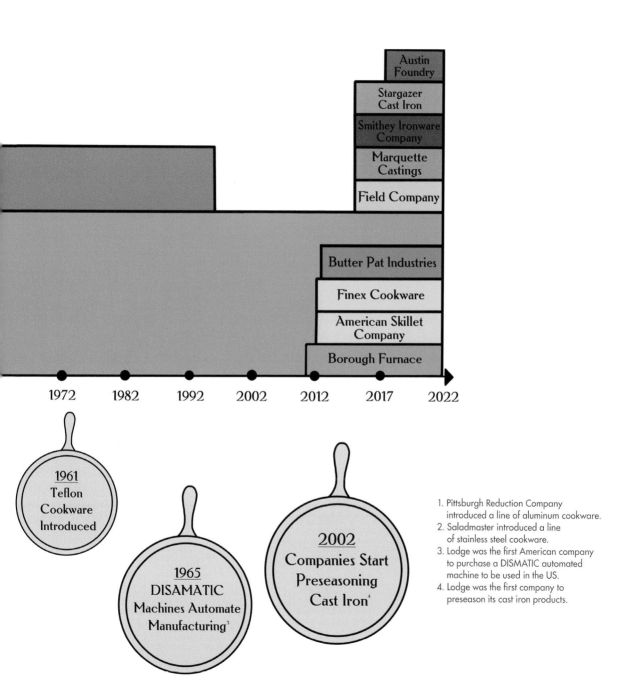

Austin Foundry

Stargazer Cast Iron

Smithey Ironware Company

Marquette Castings

Field Company

Butter Pat Industries

Finex Cookware

American Skillet Company

Borough Furnace

1972   1982   1992   2002   2012   2017   2022

1961
Teflon Cookware Introduced

1965
DISAMATIC
Machines Automate Manufacturing[3]

2002
Companies Start Preseasoning Cast Iron[4]

1. Pittsburgh Reduction Company introduced a line of aluminum cookware.
2. Saladmaster introduced a line of stainless steel cookware.
3. Lodge was the first American company to purchase a DISMATIC automated machine to be used in the US.
4. Lodge was the first company to preseason its cast iron products.

Ensuring the cast iron is smooth. *Courtesy Marquette Castings.*

Collecting Modern Cast Iron

Something interesting happened between 2010 and 2015: over a dozen individuals from around the United States all had the same epiphany. They compared their vintage cast iron to the modern equivalent and realized there was a vast difference in the workmanship. Where the vintage pans were handcrafted, smooth, and lightweight, the modern pans were mass-produced, highly textured, and relatively heavy.

These folks all came to the same conclusion: there was a wide-open market for "modern vintage" cast iron . . . and they were up for the challenge.

Thus began the Craft Cookware Movement that spawned nearly a dozen different manufacturers, each with its own take on cast-iron cookware. You'll find most of these companies outlined in the following pages. A lot can be said about each company, but I've focused on how they came about and what makes their cast iron unique. I've also included such facts as the price and weight of each company's 10" skillet (or their equivalent) to help you compare the different companies and find the piece that works best for you.

Smithey skillets. *Courtesy Smithey Ironware.*

Shopping for Modern Cast Iron

BENEFITS

With a trove of vintage cast iron still on the market, why buy a new pan? One of the biggest benefits is that you know what you're getting—and what you're not getting. Obviously, there should be no cracks or noticeable dents in the pan. Modern companies provide excellent warranties, but make note of any stipulations. Return the pan if you do find any defect because the heating and cooling of cast iron can exacerbate the smallest problem.

With new pans, you can also determine exactly what seasoning the manufacturer used. Contrary to some online sources, there is no need to strip and re-season the pan if proper edible oils are used. Simply rinse or wash the pan before use, then use a liberal amount of oil or fat when you cook.

Peter Huntley of Stargazer is packing skillets for shipment. *Courtesy Stargazer.*

WHERE TO SHOP

- Retail stores like Target and Walmart often carry Lodge Cast Iron because it is the most economical of the modern brands.

- Kitchen stores typically carry at least one or two brands of cast iron.

- Restaurants like Cracker Barrel often carry some.

- Manufacturer websites are the best place to purchase cast iron online. You'll know you're getting a quality product with a warranty, and you can often find sales.

Want to hear more from the manufacturers? I recorded several of the company interviews and have posted them online. Visit my website at ashleyljones.com for a complete listing.

One place you should avoid buying cast-iron cookware is Amazon. Much of the cast iron sold on Amazon is made by foreign companies that do not use the same level of quality control as our domestic companies. That means the metals used may not be in a proper ratio or may contain contaminants that can cause hot spots in the cookware, warping, and cracking. Or the company may use chemically based seasoning instead of edible oils.

Amazon is also fraught with counterfeit products. Unscrupulous companies will steal designs from well-known manufacturers and pass them off as authentic pieces. So that high-end skillet you've been eyeing may not be the real deal.

WHAT TO LOOK FOR

Buying new cast iron is a lot simpler than buying vintage pieces. Simply find a piece that is comfortable to hold and easy on the eye and that fits your cooking needs.

WHAT TO AVOID

- Foreign-made cast iron due to poor quality.

- Cast iron associated with celebrity cooks. Most Skilletheads avoid these pans as they are often made in China or are of poor quality. Modern cast-iron manufacturers are famous enough; they don't need celebrity names attached to them.

PRICE

Vintage cast iron may have sold for $2 many years ago, but don't expect those prices with modern cast iron.

If you're on a budget, look for Lodge Cast Iron. They provide the most economical brand of American cast iron, with a 10" pan costing about $22. From there, the price jumps to $145 for a Field Company skillet and goes up to $300 for a Borough Furnace skillet. Why the disparity? Lodge focuses on efficiency and consistency. Through automation, they're able to produce thousands of skillets every day.

In contrast, new manufacturers often spend many hours handcrafting their pans, smoothing the cooking surfaces, and seasoning them in ovens. The cost of their pans reflects the man-hours involved.

Just remember: a cast-iron pan can last you several lifetimes. That's why they're referred to as "modern heirlooms." So, how much would you value a pan that you'll use every day for the next fifty years and then pass on to your kids?

> "Foreign pans often include material made from automotive blocks."
>
> —Kent Rollins

Surface Finish

All of the modern cast-iron manufacturers I interviewed (except for Lodge Cast Iron) insist that a smooth cooking surface is more nonstick than a textured surface and is especially helpful when it comes to searing meats or cooking delicate foods.

Others, such as manufacturers of seasoning products, claim that it's actually the quality of the seasoning that determines whether a pan is nonstick.

So what really determines a pan's nonstick ability? It's most likely a combination of surface finish and good seasoning.

Peter Huntley, founder of Stargazer, described the relationship between texture and seasoning on a minuscule scale. If the surface of the pan is machined completely smooth, he said, the seasoning will slide right off while cooking. If the pan has a lot of texture, it will hold the seasoning well. However, deep grooves in the texture cannot be filled completely with seasoning, so food will likely stick to the exposed peaks of the textured pan.

Peter said they wanted to produce smooth pans, but they needed the seasoning to stick (and not the food). Their solution was to machine the pan smooth, then add "micro-texture" back to the pan to hold the seasoning. "Now, we're machining it, grinding it, and reintroducing some texture in a proprietary

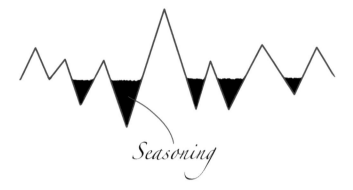

Deep Texture

Seasoning

Smooth Texture

Seasoning

Minimal Texture

Seasoning

Illustration of seasoning applied to three different surface finishes.

process to raise the surface a little bit. And then we're doing our final tumbling surface finish, which is our final smoothing. I hate to call it roughing it up, but we're reintroducing some texture at one of our middle steps so that seasoning has something to grab onto at the end."

Liz Seru said they also make use of a micro-texture at Borough Furnace. "We do smooth [our pans] by hand, so they're smooth but not slick. I always say they're soft to the touch. We specifically put a finish on it that helps it grab the seasoning to make it more conducive to home seasoning without having to necessarily be a cast iron expert to build up a really good seasoning."

Those who laud a smooth-surface finish are often referring to the texture of the cooking surface of the pan. After all, that's where meats are seared and eggs are flipped. However, the inner walls of the pan must also be considered. The main issue with the walls, though, is not that food sticks to them but that the seasoning there can be more easily "eaten away" during the cooking process.

Michael Griffin, brand director for FINEX, said they addressed this problem by leaving a "natural pebble finish" on the interior surface walls for the seasoning to cling to. He said this finish "tends to hold the seasoning a little bit better when there's more scraping and bubbling. So if you're making a tomato sauce, the bubbling of the sauce around the edges is going to take the seasoning down a bit. So having some pebbling there helps it stick."

With so much cast iron on the market today, surface finish is one of the biggest differentiators between companies. For them, it's a matter of trial and error that, as Peter Huntley said, is "one of the most difficult parts of the process."

For you and me, surface finish is like every other aspect of cast iron—a personal preference.

"There's a Goldilocks zone. People think that smoother is better and that lighter is better but it's actually quite a bit more complicated than that in both cases."

—Peter Huntley, Stargazer

"You can get almost ANY skillet to cook well! It's more about heat control and following directions. But in my opinion, it is a bit easier to start off with a silky smooth surface from the beginning, especially for first-timers!"

—Cast Iron Kev

Molds full of hot iron. *Courtesy Austin Foundry Cookware.*

Modern Terminology

CASTING METHODS

By definition, cast iron is cast, or poured into molds, to create the desired shape. Sand casting is considered the most traditional and cost-effective method, but it doesn't always produce the desired effect. Depending on the exact method and type of sand used, it can result in a textured surface that some companies then grind down.

As Sean Girdaukas of Austin Foundry told me, "There is no manual on how to produce cast-iron cookware." Each company must experiment with the options available and decide what works best for them. Interestingly, American Skillet Company chose to use the core casting method while Marquette Castings uses investment, or lost wax, casting. Most other companies, including Field Company, still use sand casting.

When I asked Stephen Muscarella, cofounder of Field Company, about their manufacturing process, he said all of the methods available today "have the same amount of pain in the ass as I had to deal with to get where I got. So I think we're happy where we're at with our process."

CNC MACHINING

Vintage pans were often hand-milled to remove excess texture and rough spots that occurred during the molding process. Collectors highly value these pieces but note that the thickness and weight vary from piece to piece based on how much iron was removed during the hand-milling process. Hot spots can also occur within the pan if too much iron was removed in one area.

While modern cast-iron manufacturers often shun automation, they recognize the drawbacks of hand-milling and utilize CNC machining, instead. Computer numerical control, or CNC, is defined as "a subtractive manufacturing process that typically employs computerized controls and machine tools to remove layers of material from a stock piece—known as the blank or workpiece—and produces a custom-designed part."[1]

In the creation of cast-iron cookware, manufacturers use CNC machining for two purposes: (1) to produce a smooth-surface finish and (2) to create the desired thickness throughout the pan.

HEAT RING

Some modern manufacturers still use heat rings, or machining rings, to ensure their pans are level. These traditional-style heat rings protrude from the pan, causing the pan to lift slightly from the heat source.

Other manufacturers create an incised or grooved ring on the bottom of the pan. While this gives a nod to the vintage heat ring, it is only for aesthetic purposes.[2]

POLISHED

Unlike CNC machining or hand-milling, polishing is not a subtractive process. Dennis Powell of Butter Pat Industries explained the difference between their polishing process and the milling that other companies perform: "What we are doing is forcing the surface of the iron to 'lay down.' We are peening the surface into a flatter shape rather than milling and cutting that roughly cast surface off to make it lighter and less rough."

PRE-SEASONED

In 2002, Lodge Cast Iron became the first company to pre-season their pans in the factory. Now, all modern cast iron is pre-seasoned, which allows you to cook on your new skillet as soon as you get it home.

   Every manufacturer uses a different oil or blend for their seasoning. This results in color variations among modern brands, from bronze and gold to brown and black. As you cook in your skillet, additional seasoning will be created, thus turning your pan darker over time.

WARRANTIES

Every cast-iron manufacturer I interviewed provides at least a limited warranty, with most of them offering a full lifetime warranty. Before you buy, check what kind of warranty the manufacturer offers, making note of any terms they stipulate.

Manufacturers

AMERICAN SKILLET COMPANY

Website americanskilletcompany.com
Founded by Alisa Toninato and Andrew McManigal in 2012
Manufactured in Madison, Wisconsin
Seasoned with a mixture of flaxseed and canola oil
Price for USA skillet $130
Weight of USA skillet 3.5–5.5 lbs

Catalog Seven skillets shaped like US states and a US-shaped skillet
"Art of the State Cookware"
Identification marks on skillets
- Bottom of pan is incised with "American Skillet Company"
- US- or state-shaped pan
- Traditional black color out of the box
- No helper handle
- No pour spouts
- No heat ring
- Relatively smooth surface all over

Features advertised
- The original state-shaped pans
- Ergonomic design
- Sized to fit on stove burners
- The core casting method allows the pan to be relatively smooth without the need for hand-milling

USA Skillet. *Courtesy American Skillet Company.*

Every manufacturing company has a unique story of how it began. However, the pans from American Skillet Company are the only ones that started out as award-winning art pieces.

Alisa Toninato has a background in fine art sculpture and foundry practices. She loves the way cast iron can make beautiful works of art, but she turned to cookware early in her career as a means of sharing useful art with others.

As soon as Alisa graduated from school, she built her own studio with a furnace. The first few molds she made were for the Wisconsin skillet (her home state). During our call, Alisa admitted, "It was such an off-the-cuff idea. . . . I just took the vehicle of cookware and state pride, threw them together and put a handle on it."

Alisa Toninato with her cast-iron skillet map on the *Martha Stewart Show.*

There was a lot of interest in the Wisconsin skillet, so Alisa designed a map of the United States with each state comprised of an interlocking skillet. After getting help from local artists and proud supporters, Alisa and her life and business partner, Andrew McManigal, debuted the 500-pound, 9' × 6' art installation in the 2011 Art Prize Exhibition in Grand Rapids, Michigan. That led to an appearance on the *Martha Stewart Show* and a special award from Martha Stewart.

"I love cookware as a tool to gather people."

—Alisa Toninato

It wasn't long before individuals began requesting their own state pans. Excited to move her art into industry, Alisa and Andrew founded American Skillet Company in 2012. Unable to keep up with the demand in her own studio, Alisa leased space in existing foundries and worked with skilled craftsmen to reimagine the state skillet concept.

Alisa explained to me what was involved in transforming her designs into skillets that could be batch processed. "There's a lot of one-on-one that I'm allowed to spend with art pieces. But when you're in production, that just needs to work every single time without someone having to baby it along.

"So there are decisions that had to be made. The skillets needed to be moldable. They needed to be shippable, stackable. Each one needed to be ergonomic, not just while it's empty but while it's full, too. We had to think about where that center of gravity was. . . . And then just making it fit on a burner. Making it realistic for people's use. Making it realistic for a recipe in general, for the volume of food that it can hold. So there were a lot of decisions that had to be made to totally reenvision it from an art piece to a production piece."

State skillets are perfect for brunch, appetizers, and desserts—basically anything you want to share on the table. "We call it the party pan," Alisa said. "It's for gatherings. It's for people to bring a smile."

Currently, American Skillet Company offers state skillets for Illinois, Michigan, Minnesota, New York, Ohio, Oklahoma, Texas, and Wisconsin, as well as a USA skillet. If your state isn't in the lineup, you can go to the company's website to vote for it. Once enough votes are received, the company will offer preorders at a discounted price. That money is then used to turn Alisa's artwork into beautiful pans that can be batch processed . . . in Wisconsin, of course.

AUSTIN FOUNDRY COOKWARE

Manufacturer Austin Foundry
Website austinfoundrycookware.com
Founded by Sean Girdaukas and Lisa Girdaukas in 2018
Manufactured in Sheboygan, Wisconsin
Seasoned with a proprietary blend of coconut oil,
   almond oil, avocado oil, and beeswax
Price for 10" skillet $140
Weight of 10" skillet 6.1 lbs
Catalog Skillets in three sizes, the largest of which
   is 12" in diameter; 12" dual handle skillet
"Built to last a lifetime and longer."
Identification marks on skillets
   · Bottom of pan is incised with "AUSTIN FOUNDRY COOKWARE," the
     foundryman logo, the size of the pan in inches, and "SHEBOYGAN WI USA"
   · New pans are seasoned black
   · Forked handles with unique contoured shape
   · Small tab-shaped helper handle incised with "AFC" on top
   · Double pour spouts
   · Smooth interior surface
Features advertised
   · Individually molded, poured, and finished by hand in the Austin Foundry in
     small batches
   · Patented handle design offers comfortable grip and superior leverage
   · Interior is CNC machine–finished, then the entire pan is polished smooth

You could say Sean Girdaukas has cast iron in his blood. Well, not in his blood
exactly, but definitely in his family, which has owned and operated the Austin
Foundry in Sheboygan, Wisconsin, since its founding in 1946. The foundry has
always been important to Sean, who started working summers there at the
age of fifteen. Awaiting admission to medical school, he began working at the
foundry full time "and never looked back." Now the VP of Austin Foundry, Sean
told me, "I'm forever grateful that I have been able to work in our family-owned
business alongside my father and brother for many years."

Bottom of 10" pan. *Courtesy Austin Foundry.*

Pouring iron into molds. *Courtesy Austin Foundry.*

Although Austin Foundry produced all kinds of cast-iron products over time, they had never delved into the world of cookware. Sean explained, "My wife, Lisa, and I thought about this for many years. Several current brands on the market came to Austin Foundry over the years asking if we could produce their cookware. Our answer was always emphatically NO. We always had it in the back of our minds that eventually we may start our own brand of cast-iron cookware."

Finally, in 2018, Sean and Lisa took the plunge and started the Austin Foundry Cookware brand. Sean's father, the president of Austin Foundry, was on board from the beginning. "He is an excellent cook, has always been a fan of cast-iron cookware, and he was more than happy and helpful in getting our brand launched."

Although the Austin Foundry had plenty of experience molding cast iron, Sean told me, there was still a learning curve. "There is no manual on how to produce cast-iron cookware. It was a long process of trial and error, from design changes, figuring out the correct gating system (which feeds the mold), avoiding gas issues when pouring the mold, figuring out the correct seasoning process, etc. . . . Thankfully, with seventy-six years of casting knowledge we were able to figure it out."

The most unique feature of the Austin Foundry Cookware pan is probably the patented handle. It is designed to be comfortable to grip as well as provide proper leverage when you lift the pan.

"I feel what we are best known for, aside from our design, is our craftsmanship and attention to detail. Our cookware is not mass produced but is individually molded and hand poured in our own foundry, small-batch style. This allows us to control every aspect of producing our cookware."

Although every manufacturer has its own story, it turns out that owning your very own foundry comes in pretty handy when you launch your own craft brand of cast-iron cookware.

"The humbling thought of how many meals and dishes that will be prepared for decades with our cookware puts a smile on our faces."

—Sean Girdaukas

BOROUGH FURNACE

Website boroughfurnace.com
Founded by John Truex and Liz Seru in 2011
Manufactured in Owego, New York
Seasoned with organic flaxseed oil
Price for 10" skillet $300
Weight of 10" skillet 6 lbs
Catalog Skillets in two sizes, the largest of which is 10.5";
   12" braising skillet; bakeware; grill pan; and a 5.5 quart
   Dutch oven with a seasoned or enameled finish
"Timeless design, quality materials and expert craftsmanship"
Identification marks on skillets
   · Bottom of pan is incised with "BOROUGH FURNACE"
   · New pans are dark in color
   · Long handles have forked attachment, are raised and celery-shaped
   · Large helper handles with loop design
   · No pour spouts
   · Gently sloped sides
   · No heat ring
   · Smooth finish
Features advertised
   · Handcrafted and finished cast iron made in small batches
     by founders John and Liz
   · Modern, minimalist designs
   · Long handles help to balance out the weight of the skillet
   · The handles stay cool thanks to their celery shape, forked
     connection with a wide base, and long length; they're also raised
     above the pan to keep them away from the heat source
   · Made from 100 percent recycled iron

Around 2009, John Truex saw an opportunity in the market for modern cast-iron cookware. He had studied metal casting as an undergraduate, had a degree in product design, and had experience designing various household products.

Courtesy Borough Furnace.

It seemed he was just the guy to start a successful cast-iron manufacturing company.

When John met future wife and business partner Liz Seru, he was already designing the first product, a 9" skillet. Fortunately, Liz had her own unique talents to add to the company, with experience in studio arts and film production. Together, they formed Borough Furnace and approached cast-iron manufacturing as an art form.

Liz explained, "We just approached this from a different place. If you want the classic skillet look, you can get that. We wanted to try to do something that made our pieces unique to us."

Borough Furnace's skillets are definitely unique. The loop-style helper handle is large

"Because we are able to machine our own tooling and we make our own castings in-house, we prototype our products endlessly—taking them home, cooking with them, and making refinements until the details are just perfect."

—Liz Seru

John and Liz in their safety gear. *Courtesy Borough Furnace.*

FUN FACT When Borough Furnace first started, they used restaurant fryer grease to fuel their furnace. It wasn't long before biofuels became more popular and harder for them to acquire. When they began scaling up their production, they had to switch to traditional heat sources.

enough to use with an oven mitt. The walls of the skillet flare out. The handle has a celery shape and a forked attachment and is extra long and raised above the pan—all features that reduce heat transfer, keeping the handle cool to the touch during cooking. The long handle also provides a balanced weight distribution.

While most companies have teams of people involved in the manufacturing process, John and Liz perform everything themselves with the aid of just one employee, from pouring the metal into casts to seasoning them. Yet they've been able to extend their catalog by adding such products as bakeware, a griddle, and a roasting pan.

Enameled Dutch Oven. *Courtesy Borough Furnace.*

Even more impressive: Borough Furnace is now producing the only enameled Dutch oven made in the United States! And yes, they're doing the enamel work in-house. "For John and I, given that we're so in the mix on everything, and really interested in the process and the craft behind everything, we didn't want to go with an overseas company. . . . So we started doing it ourselves, and that's been a process over the past couple years to really fine-tune it and get it where we want it." Amazingly, Borough Furnace's enameled Dutch oven is less expensive than its counterpart from Le Creuset.

Although Borough Furnace has focused on scaling up and introducing new products, they are still a small foundry. However, their smaller scale and ability to prototype their products in-house has afforded them many benefits and will help them to grow well into the future.

BUTTER PAT INDUSTRIES

Website butterpatindustries.com
Founded by Dennis Powell in 2013
Hand-cast in Pennsylvania
Finished and hand-seasoned in Maryland
Seasoned with beeswax and coconut oil
Price for 10" skillet $195
Weight of 10" skillet 4.8 lbs
Catalog Skillets in four sizes, the largest of which is 14" in diameter; two flat-bottom pots; glass lids (no cast-iron lids at this time)
"Inspired by the past, innovated for the future."
Identification marks on skillets

· Some pans have no markings; some have "BUTTER PAT INDUSTRIES"; and some have "BUTTER PAT INDUSTRIES CAST IRON" and "USA."
· New pans are bronze in color
· Each style and size of pan has a name, and the first initial of that name is incised on the top of the handle
· Large tab-style helper handle
· Double pour spouts
· No heat ring
· Smooth finish on all surfaces

Features advertised

· Hand-cast in limited quantities
· Thinner side walls make the pans lighter weight while a thicker heat plate provides even heating
· Smooth finish on all surfaces thanks to a unique molding process (patent pending)
· Handles are designed to conform to the natural angle of your arm when picking up a pan, placing less stress on the wrist
· The large helper handle encourages the use of both hands

Looking back, it seems Dennis Powell was destined to start a cast-iron manufacturing company. He shared his story with me.

Helper handles on Butter Pat pans.
Courtesy Butter Pat Industries.

"In the late '70s I owned a hardware store while in college. We were Lodge dealers. One day a man asked me if I knew about the smooth cast iron of the late 1800s to early 1900s, especially that from the Favorite Stove and Range Company. I was studying studio fine art at the time, including sculpture and casting. Learning about these early foundries began an academic interest in the history of cast iron, not just skillets but building facades, stoves, machinery, and even jewelry.

Handles on Butter Pat pans.
Courtesy Butter Pat Industries.

FUN FACT Each of the Butter Pat pans are named after important women in Dennis Powell's family: Lili, Joan, Heather, Estee, and Aunt Alma. The Dutch ovens, on the other hand, are named after men: Homer and Eric.

"After art school I studied architecture and for twenty-five years owned a construction company that specialized in historic renovations. I designed and built hundreds of kitchens in new and adapted spaces. This was critical training in the ergonomics of kitchen tools, adaptive reuse, and learning the lesson that the designers before you might have already done a better job."

That lesson took on a new importance when Dennis cracked his late grandmother's cast-iron skillet. Dennis's grandmother, Estee Hilton Rudd, was a true inspiration to their family, owning her own meat market in Charleston, South Carolina, in the 1920s and '30s. Dennis wanted to pass down his grandmother's legacy—in the form of her cast-iron pan—to his own sons, Roan and Grayson.

At first, Dennis set about to repair his grandmother's skillet, but he soon learned that was impossible. Then he decided to make new skillets for his sons and quickly became engrossed in the study of cast-iron manufacturing. "I visited twenty-two foundries, studied the old patents, holed up with the original Griswold factory records, cast hundreds of molds, and, two years later, still didn't have a model that met my specifications." And given his background, Dennis's specifications were really specific: (1) the as-cast wall thickness had to be less than 0.09375"; (2) the as-cast roughness average (or the measure of smoothness) had to be less than 90; and (3) it had to be made in the United States.

As it turned out, this was quite an ambitious list, one that proved impossible given existing methodologies. So Dennis and his team developed entirely new casting technologies to produce their cookware on their own terms.

Finally, Dennis knew he could make the pans of his dreams for his children, but why stop there? Perhaps he could put his knowledge to work and start manufacturing pans commercially. That's when Butter Pat Industries was born.

Dennis's sons now have skillets they can be proud of, ones that were manufactured by their own father. As for Estee's legacy, she has her own skillet now: the Estee 8" Butter Pat pan.

"Butter Pat Industries is a design-oriented manufacturer of cast-iron cookware that believes beauty can still be part of utility."

—Dennis Powell

FIELD COMPANY

Website fieldcompany.com

Founded by Christopher and Stephen Muscarella in 2015

Manufactured in Indiana, Illinois, and Wisconsin
  (with home office in Brooklyn, New York)

Seasoned with grapeseed oil

Price for 10.25" skillet $145

Weight of 10.25" skillet 4.5 lb

Catalog Skillets in five sizes, the largest of which is a little over 13"
  (Numbers 4, 6, 8, 10, 12) with lids; griddle; 4.5 quart Dutch oven

"Lighter, smoother cast iron cookware"

Identification marks on skillets

- Bottom of pan is incised with "FIELD COMPANY," the Field
  logo, "MADE IN USA," and the number of the pan (e.g., "8")
- Tab helper handle
- No pour spouts
- Heat ring
- Smooth cooking surface

Features advertised

- Smoother, lighter cast-iron pan reminiscent of the best
  vintage American skillets
- Light enough for everyday cooking
- Smooth, naturally nonstick cooking surface
- The numbering convention matches traditional vintage sizing
- Uses green sand castings like Griswold and Wagner did

Mrs. Muscarella had already replaced her cast iron with modern Teflon pans when sons Stephen and Christopher were growing up. But when the boys went off to college, she found her old cast iron and passed it down to them.

Stephen received a small block Griswold #10 and three-notch Lodge #7. Christopher received a late 1950s Wagner #8.

Years later, Christopher—who enjoyed cooking—bought a modern cast-iron pan and remarked that it wasn't of the same quality as the old pans his mother

Field Company pan. *Courtesy Field Company.*

Pouring iron into molds. *Courtesy Field Company.*

"Our pans have a classical sensibility with just enough sexiness to keep you interested."

—Stephen Muscarella

had used. By that time, Stephen was a craftsman, building furniture, doing leatherwork, and learning to appreciate high-quality tools, including cast iron. Stephen told me, "When you get really into cooking and you start wanting to dial in your gear, you kind of want to have a nice version of something. And I said, 'Well, we could make 'em.'" Christopher, the entrepreneur in the family, jumped at the idea.

The brothers knew from the beginning they wanted to make cast iron in a vintage style. "Cast-iron cookware has gone through two to three centuries of its current iterations to arrive at its current form. And I don't think you can argue with that too, too much. And our thinking was to not mess up what was already so amazing and to only tweak what we thought could use improvement."

That's why the brothers started with the general pattern of a vintage pan and decided on green sand casting, which is the most used and most cost-effective method of manufacturing cast iron.

FUN FACT The brothers didn't think Muscarella would make a good name for their cast-iron company, so they agreed on Field, another family name.

From there, the Muscarellas lengthened the handle a bit to provide more leverage and made it more ergonomic. They removed the pour spouts in lieu of a slightly tapered rim. Stephen explained that pour spouts in "modern manufacturing would have added a lot of cost and creates two different surface finishes. It's really not the pour spout, it's the top of the rim that matters."

They also lightened the pan. Stephen explained to me how important this change was. "You never reach for the heavy pan. The heavy pan just sits around somewhere. . . . The one you grab every single time is the one that feels right in your hand."

After all the research, planning, and tweaking, how did the Muscarella brothers know when they had developed a truly good pan? "We stopped reaching for the Wagner and started reaching for ours."

The Field Company website now claims their pan "is the closest living relative to vintage cast iron." I don't see anyone arguing that fact.

FINEX COOKWARE

Manufacturer Lodge Cast Iron

Website finexusa.com

Founded by Mike Whitehead and Ron Khormaei in 2012

Manufactured in Portland, Oregon (using Lodge foundry in South Pittsburg, Tennessee)

Seasoned with organic flaxseed oil

Price for 10" skillet $200

Weight of 10" skillet 6.3 lbs

Catalog Skillets in three sizes, the largest of which is 12", with lids; 14" double-handled skillet; grills and griddles; 5 quart Dutch oven

"Cast iron cookware for those who believe details make the difference."

Identification marks on skillets

· Octagonal-shaped pan
· Bottom of pan is incised with "FINEX CAST IRON SKILLET," "PORTLAND, ORE. U.S.A.," the size of the skillet ("No. 10"), and the FINEX logo
· New pans have a bronze color
· Stainless steel spring handle
· Tab-shaped helper handle is incised with the FINEX logo
· Eight pour spouts
· No heat ring
· Smooth cooking surface

Features advertised

· Eight rounded corners serve as pour spouts that seal up tight or serve as evenly spaced vents when using the matching lid
· Eight flat sides make it easier to use a spatula to remove baked goods
· The unique spring handle is wound from 300-series stainless steel rod stock, which keeps the handle cooler longer
· Pebble finish texture on inside walls holds seasoning longer
· CNC-machined smooth cooking surface for nonstick cooking
· Thick base and walls distribute heat better and provide even cooking

8" FINEX Skillet with Lid. *Courtesy FINEX.*

Assembling skillet cap. *Courtesy FINEX.*

"Get the cookware that's going to help you cook better and inspire you to do so."

—Michael Griffin

You've heard the saying "think outside of the box." Well, FINEX decided to think outside of the circular pan and created an octagonal-shaped skillet.

I spoke recently with Michael Griffin, brand director for FINEX. He helped with the company's branding from the very beginning, when Mike Whitehead and Ron Khormaei founded the company in 2012 and Lodge was the only cast-iron manufacturer on the market. Mike and Ron are no longer with FINEX, but Michael was able to describe the thought back then: "We gotta spark a new conversation because nothing has changed in one hundred years. So let's not shy away from making something visually different . . . sort of arresting."

Michael explained, "We want the design to be visually inspiring and appealing . . . but we also want that design to add some functions and features that enable people to cook better." This is what FINEX calls *premium functional design*.

From the very beginning, FINEX worked with chefs, bakers, and cast iron aficionados to try to solve the problems they were experiencing with traditional cast-iron products. They determined that it would be easier to pour if there was a spout closer to the handle and if the sides curved toward the pour spouts. Since the first slice of cornbread usually gets messed up by using a flat spatula, a flat side should help with dishing out baked goods. Then there was the hot handle to contend with. Manufacturers of vintage wood stoves and waffle makers solved this problem by wrapping the handle with a metal coil.

FINEX took all of these ideas into account and developed a pan with a round bottom, eight flat sides, and a stainless steel spring coil handle.

FUN FACT What does the name FINEX mean? Michael Griffin said it goes back to their goal of manufacturing something inspiring and functional that their customers can make their own. "We provide the 'Fine'—the refined design and production process. Our customers are the 'X factor.' They're going to bring their own family recipes, their own dreams, and their own cooking traditions. And we're happy to be a part of that."

As for the smooth versus textured question, FINEX takes a combination approach. They use a unique stone tumble polish to ensure their pans are smooth enough for use on glass-top cook surfaces. They then use CNC machining to create a smooth cooking surface while leaving the interior walls with a natural pebble finish to help hold the seasoning.

Making an octagonal pan with a unique handle requires true craftsmanship. This is one reason why FINEX agreed to an acquisition by Lodge in 2019. Michael explained that Lodge's extensive experience and quality control have helped FINEX reach the consistent high-quality standards they needed. Fortunately, with the support of Lodge, we're sure to see FINEX around for a long time to come.

LODGE CAST IRON

Website lodgecastiron.com
Founded by Joseph Lodge in 1896
Manufactured in South Pittsburg, Tennessee
Seasoned with natural vegetable oil
Price for 10.25" skillet $21.95
Weight of 10.25" skillet 5.35 lbs
Catalog Skillets in nine sizes, the largest of which is 15"; bakeware;
grills and griddles; outdoor cookware; heat-treated cast iron;
carbon steel products; enamelware

"Creating heirloom-quality cast iron cookware in Tennessee since 1896."

Identification marks on modern skillets
- Bottom of pan is incised with "LODGE"
- Seasoned black
- The teardrop handle has stayed the same over time
- Loop-style helper handles on bigger pieces
- Newer pieces have "Lodge" incised on the lid and the top of some helper handles
- Pour spouts
- No heat ring
- Textured finish

Discontinued brands include
- Chef's Choice
- Friar's Friend
- Maid of Honor
- Plantation Ware

Current brands
- Lodge Cast Iron ("Classic")
- Chef Collection
- Blacklock
- FINEX

Features advertised
- Textured cooking surface helps the seasoning adhere to the pan
- Unparalleled heat retention and even heating

Classic Lodge skillet. *Courtesy Lodge Cast Iron.*

· Seasoned with an easy-release finish that improves with use
· Smooth, easy-to-grip handle

Lodge Cast Iron is the only manufacturer that is both vintage and modern. Joseph Lodge founded the company in 1896 under the name Blacklock. The foundry was destroyed in a fire in 1910 but was soon rebuilt under the name Lodge.

A lot has happened in the past 125 years, but Lodge has survived it all. During the Great Depression, Lodge manufactured novelty items such as garden gnomes to keep workers employed. When automation was introduced to the cookware industry, Lodge became the first manufacturer in the United States to use a DISAMATIC automated molding machine. Years later, Lodge replaced their coal-fire cupola furnaces with an electromagnetic induction melting system. In 2002, Lodge became the first company to season its cast iron at the foundry, creating a new industry standard. While still focusing on seasoned cast iron, Lodge ventured into the world of enameled cast iron in 2005. In 2017, Lodge built a new state-of-the-art foundry and, in 2021, began investing millions of dollars

Damaged cast iron ready to be recycled into new pans. *Author's photo.*

to expand and reconfigure their original foundry. It seems that Lodge's key to success is to embrace innovation and to remain flexible.

Perhaps that's why Lodge acquired FINEX Cookware in 2019. Lodge had already released Blacklock, their first line of high-end cast iron. Lightweight and triple-seasoned, Blacklock is sold through Lodge's website and Williams Sonoma stores. Still, FINEX's unique octagonal shape and machine-smoothed cooking surface is a new type of luxury cast iron for Lodge, and it lets them enter new markets under the FINEX name.

This doesn't mean Lodge is abandoning their classic designs or their budget-friendly products, though. Adam Feltman, PR and Advertising Manager for Lodge, said this: "Providing a range of cookware that lasts for generations is at the heart of Lodge. As we continue to release new cookware, we'll keep our broad customer base in mind to deliver cookware that can span a range of uses and styles."

So what does the future hold for Lodge? Adam hinted at more innovation: "We're looking forward to a future of exciting new product releases in the next several months and beyond. As we develop new products, we're always looking to fill a need for cast-iron cookware, whether it be bakeware—which Lodge released as the first full line of cast-iron bakeware in 2020—or a new spin on highly sought-after enameled cast iron."

FUN FACT
Every year, Lodge is one of the sponsors of the National Cornbread Festival, which is held in Lodge's hometown of South Pittsburg, Tennessee. Lodge usually hosts tours of their factory during this time. Visit nationalcornbread.com for details.

MARQUETTE CASTINGS

Website marquettecastings.com
Founded by Eric, Kurt, and Karl Steckling in 2015
Manufactured in Royal Oak, Michigan
Seasoned with flaxseed oil
Price for 10.5" skillet $249.95
Weight of 10.5" skillet 4.1 lbs
Catalog Skillets in two sizes, the largest of which is 13"; griddles, grill pans; enameled Dutch ovens; carbon steel products

"We make premium, best-tool-for-the-job cookware for people who, beyond necessity, want to enjoy every minute in the kitchen."

Identification marks on skillets
- Bottom of pan is incised with "MARQUETTE CASTINGS," the Marquette logo, the size of the pan in inches, and "MICHIGAN, USA"
- New pans are dark in color
- Forked handle with thin base is incised with "MARQUETTE CASTINGS"
- No helper handle
- Double pour spouts
- Flared rim
- Heat ring
- Smooth finish on all surfaces

Features advertised
- Entire surface of pan is smooth thanks to the unique investment cast method
- Lightweight pan due to thinner walls and handle
- The metal itself is more porous than other cast iron, helping it to hold onto the seasoning
- Handle design (thinner at the base, with a forked shape) keeps it cooler during cooking
- Flared rim in addition to the pour spouts helps when pouring out liquids

Eric Steckling and his wife received pots and pans as wedding presents. Five years into their marriage, the finish on their cookware started flaking. The pans

10.5" skillet. *Courtesy Marquette Castings.*

themselves were in good condition, but the deterioration of the nonstick coating made them unusable.

Although Eric hadn't grown up with cast iron, his research showed it was a durable alternative. He found a Griswold #6 on eBay and was quickly convinced of its superiority over chemical nonstick pans. But then he looked into modern cast-iron pieces: "I could not believe the difference. It boggled my mind that something could be so amazing that was sixty years old and the modern stuff was a fraction of the quality."

Skillet trees coated with ceramic. *Courtesy Marquette Castings.*

Eric and his brothers, Kurt and Karl, were always tinkerers, building everything from robots to rockets in their childhood basement. As an adult, Eric continued working with his hands. "I've always been interested in making and building things. I'm a carpenter (as a hobby) and a welder; [I do] light-metal fabrication. I've done and taught myself a number of these things. A couple years ago, I learned all about how to use a CNC router." So when Eric realized the difference between old and new cast-iron cookware, he saw an opportunity to create something new but in the style of the old manufacturers.

The Steckling brothers soon founded Marquette Castings, named after a nearby city known as a major source of iron ore. It wasn't long before reality hit, though. "Making really high-quality skillets is not easy. . . . It's low-volume. It's hard." But Eric was up to the challenge.

Instead of using the standard hand-milling technique, Marquette Castings adapted a method called *investment*, or *lost wax* casting. This method had never been used to create cast-iron cookware, probably because it's so labor-intensive. In short, aluminum tooling is used to create a detailed wax casting, which is then coated in five layers of ceramic. The wax is melted, poured out, and reused while the superheated iron is poured into the ceramic mold. The cast-iron pan then has to be removed from the ceramic encasement.

For Marquette Castings, the benefits of this unique molding process far outweigh the costs. First, it allows for designs and extreme detail that are impossible with traditional sand casting. This is how Marquette Castings was able to create such a thin base for the handle, which keeps the handle cool during cooking. Second, it means the entire surface of the pan is smooth without the need for hand-milling.

Eric now looks eagerly to the future. "How cool would it be if, thirty or forty years from now, our skillets started showing up on eBay as the ones being sought after? I feel, where we're at now with our product, that we can get there." Marquette Casting's fans agree they're well on their way.

"One of my favorite things about cast iron is the fact that it does last forever. And for me, that's just a huge incentive to put the effort into the product, and to really make it something that's amazing."

—Eric Steckling

SMITHEY IRONWARE COMPANY

Website smithey.com
Founded by Isaac Morton in April 2015
Made in Charleston, South Carolina
Seasoned with grapeseed oil
Price for 10" skillet $160
Weight of 10" skillet 5.5 lbs
Catalog Skillets in four sizes, the largest of which is 12"; 14" dual handle skillet; griddle and grill pan; two Dutch ovens; carbon steel products

"Our goal is to design and manufacture premium cookware in a vintage style for people who love to cook and people who appreciate fine craftsmanship."

Identification marks on skillets

· Bottom of pan is incised with "SMITHEY IRONWARE COMPANY," "US MADE," and the company's logo of the California quail. The skillet number is also incised; it corresponds to the size of the pan (e.g., "10" is a No. 10 pan and is 10"). "CHARLESTON SC" is incised on some pans.
· The quail logo is incised on the top of the handle for most pans
· New pans have a copper color
· Handle has a small hole
· Tab helper handle has three circular holes
· Double pour spouts on some skillets
· Heat ring
· Smooth cooking surface

Features advertised

· Satin-smooth, polished finish
· Ergonomic handle
· Holes on both handles for hanging
· Pour spouts on traditional skillets
· Lengthened handle on chef skillets
· Griddles double as lids
· Engraving options available

FUN FACT
Smithey's griddles double as lids for their skillets.

SMITHEY

IRONWARE
—
COMPANY

Seasoning pans. *Courtesy Smithey Ironware.*

How do you start a cast-iron manufacturing company from the ground up? That was my question to Isaac Morton, founder of Smithey Ironware. He said he started out by restoring cast-iron cookware for friends and family. "Everybody loved them, so I got into it. I got the Blue [and Red] Books—Griswold and Wagner books—and started to learn when the different pieces were made, what made them special . . . because I just love their finished, polished surfaces. I thought they were great pieces of cookware." After doing a lot of research, Isaac realized, "There's a lot more to vintage cookware than I thought . . . and there's an opportunity to make modern cookware in a vintage style. So that's what brought me to founding Smithey, more than anything."

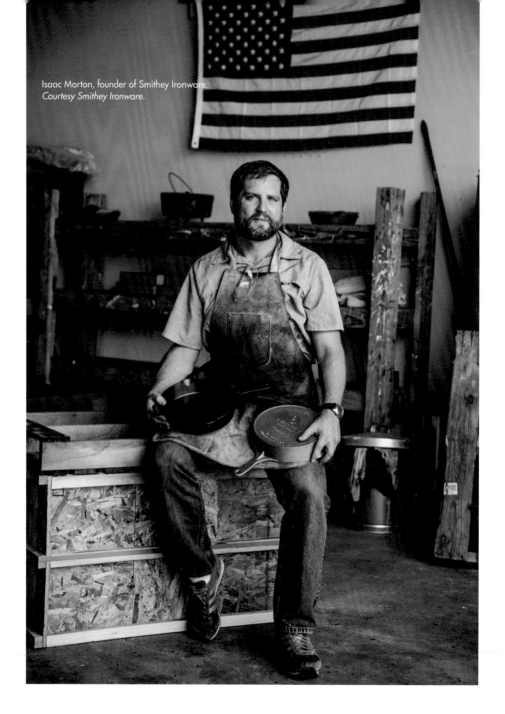

Isaac Morton, founder of Smithey Ironware.
Courtesy Smithey Ironware.

So what was Isaac's education and career before starting a cast-iron manufacturing company? Certainly, it must have been in the realm of manufacturing or art, right? Wrong. Isaac holds a business degree and has worked in the banking and finance industry. For him, cast iron was an interest, a hobby that he was able to turn into a business.

Today, Smithey is still considered a rather small manufacturer, but no one is complaining about that. Smithey boasts that each cast-iron piece is personally handled and inspected multiple times. You can even have a cast-iron piece engraved for a special occasion.

To get that smooth vintage feel, Isaac incorporated a polishing process at Smithey that is very similar to what vintage manufacturers performed. Isaac provided an overview of this multistep process: "You're taking what is a raw casting, and you're removing metal, you're grinding metal, you're blending that metal to create a surface that is naturally nonstick or very smooth." Once the seasoning is applied to the smooth surface, it creates a copper-like hue on the pans that's become a trademark of the Smithey brand. The pans will eventually darken after you use them, but the craftsmanship will remain.

What's next for Smithey? Isaac said he wants to continue to add to Smithey's cast iron and carbon steel lines. He couldn't share the details with me yet, but he hinted that he has over five different designs he'd like to add in the future.

> "Cast iron lasts forever, it's versatile, it carries an heirloom quality where it's passed down from generation to generation. So it's a very special piece of cookware."
>
> —Isaac Morton

STARGAZER

Website stargazercastiron.com
Founded by Peter Huntley in 2015
Cast in Wisconsin
Machined in Ohio
**Hand-smoothed, seasoned, and shipped from
   their headquarters in** Lehigh Valley, Pennsylvania
Seasoned with canola, grapeseed, and sunflower oil
   (unseasoned pans are available)
Price for 10.5" skillet $115
Weight of 10.5" skillet 5.2 lbs
Catalog 10.5" and 12" pans; 13.5" dual-handled braiser
**"Smooth, American-made cast-iron cookware built for
   performance, not nostalgia. We're moving cast iron forward."**
Identification marks on skillets

· Bottom of pan is incised with "STARGAZER CAST IRON," "MADE
  IN USA," the size of the pan in inches and centimeters, the
  company's eight-pointed star logo, and the pattern number
· New pans have a bronze color
· Long, forked handle that is tapered up on the sides
· Wide helper handle with a four-finger opening
· Helper handle is incised with the star logo on the top and the casting date
  (YYYYMMDD) on the bottom
· Flared rim instead of pour spouts
· No heat ring
· Smooth interior

Features advertised

· The entire interior of the pan is machined smooth and then a proprietary
  micro-texture surface finish is added to help hold the seasoning
· The weight is optimized for the perfect balance of heat retention and
  ease of use
· Unique, lifted forked handle stays cooler longer
· Drip-free flared rim (instead of traditional pour spouts) allows you to pour
  from anywhere
· The only direct-to-consumer cast-iron company on the market,
  which allows for lower pricing

Peter Huntley of Stargazer. *Courtesy Stargazer.*

My first question to Peter Huntley, founder of Stargazer, was regarding the celestial name of the company. Peter's response: he knew he wanted the name of his company to be personal, and he's always loved science and astronomy. When he thought of the name Stargazer, it seemed "timeless," a quality that cast-iron cookware certainly possesses. So that's why Stargazer Cast Iron has such a unique name and why the eight-pointed star is its logo.

Stargazer also has a unique beginning. While many of the new cast-iron manufacturers were founded by cast iron fans, Peter didn't grow up in a home that used cast iron. Instead, his background was more technical in nature. He

Polishing a skillet. *Courtesy Stargazer.*

went to school for art and design, eventually designing cookware, including ceramics and glassware. His experience helped him see the whole process, from design to production and getting the product to stores.

After a while, though, Peter became disillusioned with the cookware marketplace, especially with the way products were outsourced to other countries. "As soon as things get outsourced, corners are cut, costs are cut," he said. He knew if he ever had his own company, his products would be fully American-made.

When Peter was finally introduced to cast-iron cookware, he heard that vintage pans were made better than modern ones. Peter started restoring and using vintage cast iron and decided for himself that this was true. That's when he saw an opportunity in the marketplace to make smooth, quality cast iron once again.

This is where Peter's background in cookware design would come to the forefront. "I was excited about the vintage designs, but it was also an opportunity to update a design that hasn't been touched in fifty, sixty, seventy years. We had some fresh ideas that could be brought into that, and I think we've been able to do that."

I spoke extensively with Peter about the many features he's built into Stargazer's pans, including the "optimal weight" of each pan, the flared rim that allows you to pour from any angle, and the stay-cool handle that actually stays pretty cool. Based on Stargazer's reviews, it's clear that Peter's fresh ideas have culminated in a modern cast-iron pan with a timeless quality.

So what's next for Stargazer? Lids! And once they have lids, Peter assured me, they will be able to pursue a Dutch oven design as well. Stargazer's fans couldn't be happier.

"It was important to me to keep it American made. So that was part of the original goal: to be American made, direct to consumer, and modern design."

—Peter Huntley

FUN FACT It takes six weeks to manufacture a Stargazer skillet, from beginning to end. That's because the various aspects of the manufacture, from casting to machining to shipping, are performed in different cities across two states, and it takes time to ship the pans from site to site.

Photo by Courtney Wahl.

Restoring Cast Iron

Cast iron restoration is not a brute-force activity. Cast iron may be strong, but it's also brittle and can crack, warp, or pit if mistreated, and there's no way to fix a damaged pan.

To safeguard the cast iron, restorers have developed several processes that gently remove old seasoning, crud, and rust. Then they re-season the pans to bring them back to their former glory.

Some techniques will remove the rust but not the seasoning while others remove the seasoning but not the rust. Before you start any method, be sure it's appropriate for your needs.

The (Not So Black) Black Pan

You may think of cast iron as the "black pan" of the kitchen, but unseasoned cast iron is actually silver. It's the seasoning that turns it black.

"I'd say most restorers use the same tools and techniques, but we all have our secrets."

—Orphaned Iron

And what is seasoning exactly? It's oil or fat that's heated on cast iron, creating a polymer layer that is bonded to the iron itself. This protective layer turns dark during the heating (or curing) process.

Don't Do This at Home (Or Maybe You Can)

Experts don't always agree on how to restore cast iron, but they do agree on what to *avoid*.

Fire The old-school method of restoring cast iron was to throw it in a fire and come back for it later. The heat would remove any rust and seasoning, letting you start from scratch. People still recommend using fire today, especially as a last resort. However, if you're not careful, the extreme heat can weaken or warp the pan, reducing its ability to cook evenly. Cast Iron Savannah claims that extreme heat will also "alter the molecular structure of the metal, making it difficult, if not impossible, to season."

Pieces that have been damaged by fire often have a reddish hue, like rust but a deeper shade of pink or red.

Self-cleaning oven Many people recommend baking off the rust and seasoning in a superheated oven or during the oven's self-cleaning cycle. This presents the same problems as an open fire and also emits large amounts of fumes into the kitchen, which can be hazardous to breathe. There are even accounts of pans catching on fire thanks to baked-on crud. So while this method is largely popular online, it's a hard pass for the professionals.

Power tools I have used a power drill with a wire brush attachment a couple times to remove stubborn rust. When Skilletheads saw the video, they cringed and sent me messages asking me to never do that again. Why? Sanding, grinding, and wire wheels can all leave permanent marks on the surface of a pan, which reduces its value to collectors.

If power tools were used on your pan in a previous restoration attempt, or if you have an inexpensive pan, the appreciative value may be inconsequential. If that's you, then don't feel bad about grabbing your power drill if you need help with some stubborn rust. However, there are better methods to remove seasoning and rust, as outlined in this section.

If you don't want to restore your own pan at all, no problem! Skip to "Meet the Skilletheads," to find someone to do it for you.

Are Chemicals an Issue?

Cast iron restorers will often turn to products like Easy-Off, Carbon-Off!, and Evapo-Rust when faced with stubborn rust or crud. Concerned about the safety of these products, I immediately discounted them. However, according to Sam Rosolina, who holds a PhD in analytical chemistry from University of Tennessee in Knoxville, these products appear to be rather safe.

"Do some test runs on newer lodge pans or Asian-made pans before attempting high-end collectibles."

—Orphaned Iron

Sam researched the chemical components that are listed in the products' Material Safety Data Sheets. He found the main ingredients in Easy-Off and Carbon-Off! are soluble in water and should rinse off cleanly after use. Sam's conclusion: "Personally, I don't see a problem using something like Easy-Off in the same way someone would use oven cleaner—finishing by heating it up in the oven" at 450°F.

Unfortunately, Evapo-Rust uses "proprietary ingredients," so Sam wasn't able to provide the same assurances. However, the manufacturer does claim their product is nontoxic, water-based, and safe for use on cookware.

This is all really good news for restorers who rely on these products for one-off restorations. It's also good news for you if you purchase restored cast iron, because one of these products may have been used during the restoration process.

To read Sam's detailed responses, with reference links, visit my website at ashleyljones.com.

Remove the Seasoning

A while back, I restored my mother-in-law's camp Dutch oven using a Vinegar Bath and a power drill with a steel attachment. Then I seasoned the pot several

Dutch oven with mottled seasoning. *Author's photo.*

times using flaxseed oil. The result was a rust-free, fully functional Dutch oven. However, it wasn't very pretty. That's because a lot of the old seasoning was still present, and it had a darker appearance than the fresh seasoning. This resulted in a weird, mottled appearance.

I was lucky that the seasoning on the inside of the pot was rather uniform. However, if a pot has a patchwork mixture of old and new seasoning, the difference in texture could cause food to stick.

To ensure a deep, even seasoning, you'll need to do more than just remove the rust—you'll need to remove all of the old seasoning as well. Once the cast iron is completely silver, you can re-season it, and it will look like new.

Here are some techniques the pros use to strip their pans of seasoning.

EASY-OFF METHOD

If you have just one piece of cast iron to restore, or you have a rather difficult piece like a tea kettle or humidifier, you may want to try spray-on oven cleaner, specifically the brand Easy-Off. This product works really well because the active ingredient is lye, which is very base and dissolves fats.

However, lye can also dissolve your skin, so it is extremely important to follow the safety guidelines on the packaging, including wearing appropriate gloves and eye protection. To avoid breathing the fumes, I also recommend wearing a respirator mask and keeping this project outside.

PRO TIP Sam Rosolina, with a PhD in analytical chemistry from University of Tennessee in Knoxville, gives the following suggestion for removing Easy-Off while avoiding flash rust: skip the water and rinse the pan two to three times, alternating between ethanol and acetone, making sure the final rinse is ethanol. Then finish by heating the pan in an oven at 450°F to remove any remaining chemicals.

Supplies

- Heavy-duty chemical-resistant gloves
- Eye protection
- Face mask
- Oven cleaner (yellow cap Easy-Off is recommended)
- Heavy-duty garbage bags
- Silicone scraper
- Steel wool pad or brush
- Water

Instructions

- Set up your supplies outside and put on your protective gear.
- Being mindful of the wind (so you don't get a face full of Easy-Off), spray oven cleaner on the entire pan, including the handle and lid if there is one.
- Place the wet pan in a garbage bag and seal it. Store in a warm place for at least one day. (You can leave the bag outside as long it's in a safe place away from curious kids and animals.)
- Transport the bagged pan to a sink or outside spigot.
- Using gloves, remove the pan from the bag. Use the scraper to remove large clumps of dissolved seasoning and then scrub the pan with a steel wool pad or brush. Rinse the pan completely with cold water.
- Repeat these steps as necessary until all the seasoning has been removed and the pan is silver.

Disposal

I recommend rinsing out the trash bag to ensure the chemicals are diluted. Then tie the bag and dispose of it in your regular trash.

LYE BATH

Oven cleaner is an effective tool to remove seasoning, but it may take multiple cans to remove the seasoning of just one or two pieces. Instead, most Skilletheads prefer a Lye Bath, which can be easily set up and reconstituted indefinitely.

Lye is environmentally safe if sufficiently diluted. However, it is very caustic when undiluted, making it quite dangerous to people and animals. Be sure to protect your skin and eyes when working with lye and place your Lye Bath away from kids and pets.

Supplies

- Heavy-duty chemical-resistant gloves
- Eye protection
- Face mask (preferably vapor-resistant)
- Heavy-duty plastic tote with lid (large enough to hold your pan)
- A second heavy-duty plastic tote or bucket to transfer the pan from the Lye Bath to the rinsing area

- Water (enough to submerge the pan)
- 100 percent pure lye—between ⅓ and ½ lb per gallon of water (Wagner Society recommends the Red Devil brand)
- Nylon or silicone scraper
- Steel wool pad or brush

Instructions

- Find a safe area to set up the Lye Bath.
- Once your tote is where you want it, fill it with water, leaving a little room at the top.
- Put on your protective gear.
- Measure out the appropriate amount of lye.
- Slowly and carefully add the lye to the water, NOT the other way around. This is very important for safety reasons.
- The mixture will seem soapy once the lye reacts to the water.
- Carefully submerge your pan in the Lye Bath and cover the tote tightly with a lid. (You can suspend the cast iron using a coat hanger, but it's not necessary.)
- Let the pan sit in the Lye Bath for at least one day, then check it. It's common for the Lye Bath to take several days to a week for really dirty pieces. The lye will not cause pitting or rusting, so the cast iron can be left there for a while.
- Using protective gear, remove the pan from the lye and place it in a second plastic tote so you can carry it safely to a sink or water hose.
- Rinse the pan well with cold water. Use the scraper to remove large clumps of dissolved seasoning and then scrub the pan with a steel wool pad or brush.
- If the seasoning has not been fully removed, soak the pan in the Lye Bath again. Repeat as needed.

PRO TIP Some Skilletheads use a "hot" Lye Bath, meaning they use more lye than is recommended. Cast Iron Kev uses twice as much lye in his bath—two pounds of lye per five gallons of water. Regardless of the amount of lye you use, be sure to follow all safety precautions.

Disposal

The Lye Bath can be used multiple times, but after a while the water will turn black and the lye will stop working. When that happens, you can dispose of the Lye Bath or simply add more lye to reactivate the bath and use it again.

If the lye is sufficiently diluted, it can be poured down the sink or on the ground. Due to the nature of septic tanks, it is not recommended to dump large amounts of lye into a septic system.

Safety note

If lye gets on your skin, immediately remove any contaminated clothing, gently brush excess lye off the skin (if in a powder form), then flush the area exposed with running water for fifteen to sixty minutes while contacting emergency services. If swallowed, contact Poison Control immediately.

CARBON-OFF!

If you have a cast-iron pan with stubborn carbon deposits, you may want to try Carbon-Off! This is the brand name of a cleaning agent comprised of surfactants, water-based solvents, alkaline detergents, and de-rusters formulated to melt away oil deposits from hard surfaces. Like lye, this stuff can damage your skin and comes with a host of health warnings, so use it with care and follow all safety precautions.

Supplies

· Eye protection
· Heavy-duty chemical-resistant gloves
· Face mask
· Paper towels or paintbrush
· Carbon-Off! gel
· Heavy-duty plastic tote or bucket, or a tarp
· Steel wool brush
· Water

Instructions

- Find a safe area to set up your project. It will need to sit in a well-ventilated area, preferably outside, for up to a day.
- Put on your protective gear.
- Using a paper towel or paintbrush, apply a small amount of Carbon-Off! directly to the areas with stubborn carbon deposits.
- Place the pan in a plastic tote or on a tarp.
- Let the pan sit for a minimum of thirty minutes up to twenty-four hours.
- Carefully scrub the affected area with a steel brush. If deposits of carbon remain, reapply the Carbon-Off! and repeat the process.
- Once the carbon is removed, rinse thoroughly with cold water.

Disposal

Rinse off remaining chemicals thoroughly. Be careful to dilute the chemicals as concentrated lye will kill grass and plants.

Remove the Rust

Cast iron may seem indestructible, but its kryptonite is water. Just a little moisture on a pan can cause it to begin rusting overnight.

If your pan has only a bit of rust, you can easily remove it and then re-season the pan in the oven. Use a wool scouring pad to gently scrape off the rust in the affected area (not the whole pan). Rinse and dry the pan. If rust is still visible, repeat the process. This may take the affected area "down to the silver," meaning the seasoning may also be removed, exposing the raw silver cast iron underneath—that's all right.

Once all the rust has been removed, rub half a tablespoon of oil over the entire pan. Using paper towels, wipe off any excess oil. Place the pan in the oven and bake it 10 to 20 degrees under the smoke point of the oil for one hour. Turn off the heat and let the pan cool in the oven. You may need to repeat this process two or three times to reestablish the seasoning. Don't be concerned about the color of the new seasoning, however; it will most likely be much lighter than the rest of the pan.

If you take good care of your pans, this is all you need to do in terms of rust removal and seasoning. If the rust is more pervasive, though, hand-scrubbing will only reveal more rust. Power drills may be tempting, but they leave tool marks on the pan. This is why Skilletheads turn to chemicals and even electricity to remove rust from cast iron. If you want to remove rust while preserving the value of an old skillet, then you're going to need to try one of the following methods.

VINEGAR BATH

For superficial rust, try a Vinegar Bath. This is a relatively easy process, and the chemicals are harmless. However, vinegar is acidic, and it can cause pitting on the pan's surface, so you'll need to monitor the pan closely and limit the amount of time in the Vinegar Bath.

Supplies
- Large stainless steel sink or a plastic container large enough to submerge the cast iron
- Distilled white vinegar (5 percent solution)
- Water
- Steel wool pad or brush

Instructions
- Fill container with a fifty-fifty mixture of vinegar and water, sufficient to completely cover the cast iron.
- After thirty minutes, use the wool pad or brush to scrape off the rust. If the rust will not come off completely, place it back in the Vinegar Bath. Check the pan every thirty minutes to determine if the rust can be removed or if pitting has begun.
- If the rust is still present after a couple hours, you may want to try a different method, as continued exposure to the vinegar could damage the pan.

EVAPO-RUST

Every once in a while, you'll come across a piece of cast iron that is too rusty for a Vinegar Bath or an enclosed piece that doesn't work well in an E-Tank (e.g., a tea

Dutch oven in a Vinegar Bath. The oven is full of the vinegar and water mixture, and the lid is inverted on top. *Author's photo.*

kettle or humidifier). Then you may want to try Evapo-Rust. This rust remover is touted as a water-based product that works in minutes and is safe on skin and eyes. To be extra cautious, however, I recommend wearing the minimal safety equipment.

Supplies
· Eye protection
· Heavy-duty chemical-resistant gloves
· Evapo-Rust
· Heavy-duty plastic tote or bucket
· Steel wool brush
· Water

PRO TIP Cast Iron Savannah said to be sure to remove all of the Evapo-Rust or the item will "smell terrible when seasoning."

Instructions
- Find a safe area to set up your project, preferably outside.
- Put on your protective gear.
- Pour the Evapo-Rust in a plastic tote or bucket and then add the cast iron, making sure the pan is submerged.
- Let the pan soak for thirty minutes, then carefully check it. If rust remains, place the pan back in the Evapo-Rust for another thirty minutes.
- Once the rust is removed, rinse the pan thoroughly with cold water.

Disposal
Evapo-Rust can be poured down the drain in most areas.

ELECTROLYSIS TANK (E-TANK)

Water and electricity don't mix . . . except in the case of electrolysis. Skilletheads use Electrolysis Tanks (or E-Tanks) to remove rust from cast iron. Some even skip the Lye Bath and use the E-Tank to remove seasoning as well. Creating an E-Tank is a relatively safe project if you use proper care, but before we dive into the details, let's take a look at what electrolysis is and how it works.

Electrolysis has many uses, but in the case of rust removal, it involves an electrical charge that travels from one piece of metal to another through an electrolyte solution. This effects a chemical change in the metal pieces, damaging one piece (called the "sacrificial electrode") while repairing the other (e.g., your grandmother's cast-iron pan).

In electrochemical terms, electrolysis decomposes chemical compounds. Some resultant atoms are liberated as a gas (hydrogen, oxygen), while other resultant atoms are deposited as a solid on the electrodes (black sodium carbonate and rust). As for the rust on your pan, it either detaches from the surface of the cast iron or is converted into a deposit that can be easily removed. This is why electrolysis is technically a rust reduction method, not rust removal, because it reduces the hard red rust to soft black rust, which you can then remove.[1]

Since electrolysis does not cause scarring or pitting on the pan, it is the preferred method of nearly all cast iron collectors and restorers. The E-Tank does produce small amounts of hydrogen, so it must be properly ventilated. However, it does not create any poisonous chemicals or gases, and the water can be dumped out safely after use. This project does involve a small electrical

E-Tank Setup

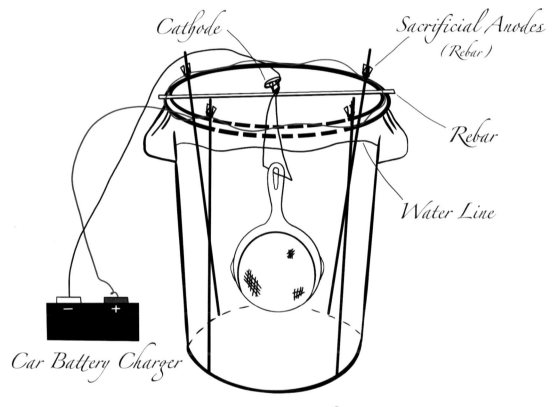

Cathode

Sacrificial Anodes
(Rebar)

Rebar

Water Line

Car Battery Charger

20 Gallon Trash Can

charge, though, so proper caution is required in the placement and tending of the E-Tank.

Now for the first step: building your own E-Tank.

Supplies

- A heavy-duty plastic container, such as a twenty-gallon garbage can. Make sure it is large enough to suspend the cast-iron pan with a little room around the edge for the anodes.
- Sacrificial anodes made of regular steel (NOT stainless steel)—Good sources include old rebar, lawn mower blades, and plain steel sheeting. Electrolysis is a line-of-sight process, so make sure you have enough anodes to cover your pan.
- Zip ties
- Wire to connect your anodes together—Ensure the wire is rated for at least double the load that your charger is rated for. Or, instead of wires, you can use a few sets of cheap jumper cables and make a series of short jumpers from them.
- Arm & Hammer Washing Soda, a.k.a. sodium carbonate (NOT baking soda, a.k.a. sodium bicarbonate)—½ cup per ten gallons of water. This will be used to create the electrolyte solution that carries the current.
- Bare steel wire, such as rebar tie wire to hang your cast iron in the tank
- A two-by-four piece of wood or an extra piece of rebar from which to hang your cast iron
- Water—Enough to submerge your cast iron
- Manual car battery charger between two and ten amps. An automatic charger will NOT work because the safety feature will cause it to shut down automatically.
- Steel wool pad or a steel wire brush for subsequent cleaning

To build the E-Tank

- Find a safe location with adequate ventilation for your E-Tank because it will produce hydrogen in small amounts. Don't put it next to your grill or burn barrel. You may also need to dump the E-Tank at some point, so if it's a large container that's difficult to move, make sure you put it in an area where you can tip it over to dump it out.

- Drill holes at the top of the container and use zip ties to loosely mount the anodes to the tank.
- Attach the anodes together using wire. A single wire should be wrapped around each anode until all the anodes are connected. (If using copper wire, be sure it doesn't come into contact with the water because it can further corrode the iron.)

To strip your pan

- Add the washing soda to the empty tank. Add water to the tank and stir until the washing soda is dissolved.
- Wrap the steel wire around the two-by-four and the other end around the handle of your cast iron.
- Rest the two-by-four across the top of the tank, allowing the pan to hang in the water. Be sure the pan does not come into contact with the anodes!
- Attach the negative (black) lead to the rebar tie wire attached to the cast iron. Do not hook these up backward! If you attach the positive (red) lead to your cast iron, it will cause it to rust and dissolve. (Remember, red will rust!)
- Attach the positive (red) lead to the sacrificial anodes.
- Turn on the battery charger and set your voltage and current to the lowest setting. The cast iron should start to bubble in the water.
- Check on the pan periodically by first turning off the power supply and lifting the cast iron out of the tank. The process is complete when all the reddish brown rust is gone or once it has turned to black. This can take from an hour to a day, depending on the amount of rust, the size and position of the anodes, and the amperage of your power supply.
- Place your cast iron in a sink and clean off any remaining residue with steel wool or a steel wire brush. If you have trouble removing the black residue, you may need to put the pan back in the tank for a while.

PRO TIP While most restorers remove old seasoning before tackling rust, Orphaned Iron takes a combined approach. He simply places the rusty, cruddy cast iron directly in the E-Tank. He explained, "Electrolysis actually removes rust, food, carbon, paint, and seasoning!"

PRO TIP Instead of removing the anodes to clean them, Orphaned Iron suspends a sacrificial piece of steel in the tank and then switches the current. This moves the rust from the anodes to the new piece of steel, thereby "cleaning" them. Then he just reverses the current again before using the E-Tank.

Disposal

The electrolyte solution will turn black after some use, but you can continue to use it indefinitely. Simply add water to replace any that evaporates. The sodium carbonate will not evaporate, so you will never need to add more.

When you want to refresh the water, simply dump the solution onto the grass.

When the anodes are dirty, they're not as efficient at removing rust from your cast iron, so be sure to clean them periodically.

VARIATIONS ON THE E-TANK

E-Tanks can be built in any size and in any plastic container sufficient to hold water. Orphaned Iron built his E-Tank out of a fifty-five-gallon plastic barrel with the top cut out of the rim. Instead of several pieces of rebar, he used sheet steel to create one big sacrificial anode that surrounds the cast iron. With complete coverage, he can use the minimum amperage on his power supply and still get quick results.

Not all E-Tanks have to be big, though. When building your E-Tank, remember that the smaller the E-Tank, the more efficient it will be. Just make it sure it's big enough to hold the largest piece of cast iron you wish to restore and there's a small space between the cast iron and the anodes.

DON'T SACRIFICE YOUR HEALTH WITH YOUR SACRIFICIAL ANODES

You may find some sites online recommending the use of stainless steel for the sacrificial anodes. Proponents state that stainless doesn't corrode like regular steel, requiring less cleanup. In fact, the stainless steel does corrode, just at a slower rate due to the chromium it contains.

The problem is that during electrolysis, the chromium [Cr(0)] corrodes and oxidizes right along with the steel, turning it into Cr(III) and then Cr(VI), known as hexavalent chromium or chromium 6. Hex-chrome, as it is commonly called, is a known carcinogen that must be legally disposed of as a hazardous waste.

Barrel E-Tank in operation. *Courtesy Orphaned Iron.*

You do not want to soak your cast-iron pan in a carcinogenic soup, nor do you want to be responsible for the safe disposal of it. But can your little E-Tank actually produce hex-chrome? I checked back with Sam Rosolina:

> I love chromium because it demonstrates the extreme difference that oxidation states can play. Cr(VI) is carcinogenic, but Cr(III) is an essential trace element for humans! The amount of Cr(VI) produced compared to Cr(III) is really dependent on the setup. If you can control the voltage in your E-Tank, you may be able to keep it below the voltage required to produce Cr(VI). In general, it's really unlikely that a home setup will produce Cr(VI) in a high enough amount to be a serious human health or environmental threat, but I say it's safest to just use a plain steel anode or a graphite carbon anode.

I agree with Sam. Don't sacrifice your own health for the sake of a skillet! When making your E-Tank, pick regular steel for your sacrificial anodes.

PUT DOWN THAT COPPER BRUSH!

Many articles warn about using copper when restoring cast iron but fail to explain why it would be problematic. Fortunately, my chemist friend Sam Rosolina was able to provide an explanation:

> Different metals have different preferences toward electrons. In the case of copper and iron, copper prefers electrons more than iron, and so, similar to oxygen, copper will steal the electrons from iron, which results in iron being oxidized (rust).
>
> In order for this to happen, there needs to be water and some kind of electrolyte (basically a water solution of charged particles like Na+, Cl–, K+, OH–, Ca^2+, etc.) so that the electrons can pass from one metal to another. Electrolytes are required in E-Tanks for the same reason, so if copper ends up in an E-Tank, or if it's used to scrub cast iron using water, it can result in this exchange of electrons which is called galvanic corrosion.

The last thing you want to do is to encourage corrosion while trying to remove rust. So if you use copper wire to connect your anodes, make sure it doesn't dip into the water. And when scrubbing your pans, choose a stainless steel scrubber instead of copper, or you could be visiting that E-Tank once again.

Final Cleaning

After you've removed the old seasoning and rust, give your pan a final scrub with a steel scrubbing pad on all surfaces. Add some mild dish detergent, like Dawn, and continue scrubbing until clean. Rinse thoroughly with cold water and dry with paper towels. If the towels become noticeably dirty, scrub the pan again with soap.

If the pan is still dirty, Skillethead John recommends wiping the pan with a Magic Eraser (the Mr. Clean brand) while rinsing frequently. Use cold water to prevent flash rust.

Once clean and towel-dried, place the pan in a warm oven (200°F–350°F) for fifteen minutes to dry thoroughly.

Now you're ready to re-season the pan.

Re-season

There's a lot of information online about the best oils and temperatures to use when seasoning a pan. To determine which method is best for you, you must first understand the science behind the seasoning.

POLYMERIZATION

During the seasoning process, oil is applied to the cast iron and then heated. The heat causes the oil to polymerize, meaning hundreds of molecules link together through chemical bonds. These polymers become trapped within the pitted surface of the pan and become partly bonded to the pan itself. This creates the nonstick layer that we call seasoning.

It's important to note that heat is required in the seasoning process, but the amount of heat is a hot topic in the cast iron community. One camp insists the oil must be heated past the smoke point

PRO TIP Crisbee just introduced a new product called Crisbee Sudz. It's currently the only dishwashing soap with oxalic acid, which prevents flash rust. You can use it after restoring your cast iron or for daily washing. Look for it on Amazon.

while the other camp argues that the oil should never be heated past the smoke point. The issue here is one of safety. That's because the smoke point is the temperature at which the oil begins to decompose and emit carcinogenic fumes. Folks like me who want to avoid breathing in these fumes will take care to season their pans below the smoke point.

Here's a list of common smoke points, courtesy of *Modern Cast Iron*.

Cooking Oils / Fats	Smoke Point	Cooking Oils / Fats	Smoke Point
Unrefined flaxseed oil	225°F	Sesame oil	410°F
Unrefined safflower oil	225°F	Cottonseed oil	420°F
Unrefined sunflower oil	225°F	Grapeseed oil	420°F
Unrefined corn oil	320°F	Virgin olive oil	420°F
Unrefined high-oleic sunflower oil	320°F	Almond oil	420°F
Extra virgin olive oil	320°F	Hazelnut oil	430°F
Unrefined peanut oil	320°F	Peanut oil	440°F
Semi-refined safflower oil	320°F	Sunflower oil	440°F
Unrefined soy oil	320°F	Refined corn oil	450°F
Unrefined walnut oil	320°F	Palm oil	450°F
Hemp seed oil	330°F	Palm kernel oil	450°F
Butter	350°F	Refined high-oleic sunflower oil	450°F
Semi-refined canola oil	350°F	Refined peanut oil	450°F
Coconut oil	350°F	Semi-refined sesame oil	450°F
Unrefined sesame oil	350°F	Refined soy oil	450°F
Semi-refined soy oil	350°F	Semi-refined sunflower oil	450°F
Vegetable shortening (Crisco)	360°F	Olive pomace oil	460°F
Lard	370°F	Extra light olive oil	468°F
Macadamia nut oil	390°F	Ghee (clarified butter)	485°F
Canola oil (expeller pressed)	400°F	Rice bran oil	490°F
Refined canola oil	400°F	Refined safflower oil	510°F
Semi-refined walnut oil	400°F	Avocado oil	520°F
High-quality extra virgin olive oil	405°F		

CARBONIZATION

Many Skilletheads insist the seasoning process works best when the pan is heated beyond the oil's smoke point. They state this is the best way to give the pan that deep, dark color, and they're right. But what they're doing is no longer polymerization—it's carbonization.

Jamie Grigg, cofounder of the BuzzyWaxx seasoning blends, explained, "If you season below the smoke point, you are not polymerizing the oils enough for them to carbonize to your iron and you'll lose your seasoning. You want to be above the smoke point of the oil used. True seasoning doesn't stop at polymerization, but carbonization."

There aren't a lot of resources online about the carbonization of cast-iron pans, so I reached out again to Sam Rosolina. He put it this way: "polymerization is when the polymer is initially formed (and carbon chains stick to the iron and line up together to form a natural plastic); carbonization is essentially burning it after it's all lined up and organized so that it's harder and more resistant to acidity, for example."

While carbonization is nice, it's not necessary for everyday cooking. Sam combined his experience—both in the lab and in the kitchen—to make an educated guess. "I think once it's polymerized that continued cooking will help the carbonization process over time; it may just not be as efficient. It probably means that re-seasoning will need to happen more often compared to those who heat it up past the smoke point during the seasoning process."

And as for the carcinogenic fumes I'm so worried about, Sam put it in realistic terms: "Anytime something burns, smoke is produced, which is inherently carcinogenic. However, with a good hood vent it shouldn't be a problem. Remember that grilling and smoking foods is also inherently carcinogenic, so it's worth adding that context and weighing those risks. Overall, as long as you're not directly breathing in the smoke for long periods of time, the health risk is very low."

TWO-STEP PROCESS

I started using cast iron when I learned that it was a healthful cookware. It infuses iron into my food, and it never leaches harmful chemicals or emits fumes like chemical nonstick cookware. So my method of seasoning still errs on the

BuzzyWaxx Original Blend. *Photo Credit: BuzzyWaxx.*

side of caution: I never season my pans above the smoke point. Does that mean it will take years for my pans to develop that beautiful dark color? Not necessarily.

Brad Stuart of Crisbee, a cast iron seasoning company, recommends a middle ground in which the polymerization and carbonization processes are handled separately. He says the oil (or any of the Crisbee seasoning blends) should be heated below the smoke point to polymerize safely. Then, the pan can be baked without oil at a high temperature, such as 500°F–525°F. Brad said that at this point, "You're basically baking the polymer on." This two-step process should give you the dark, protective seasoning you're looking for without having to worry about fumes.

Crisbee Products. *Courtesy Crisbee.*

OILS AND BLENDS

Home cooks often have a preference for the type of oil they use for seasoning, such as vegetable shortening (Crisco) or plain olive oil. Our grandmothers used lard to season their pans because that was readily available, but lard can become rancid over time and is not often used today.

The folks at the Cook's Country Test Kitchen experimented with various oils and determined that flaxseed oil creates the strongest nonstick coating due to its high level of omega-3 fatty acids.[2] However, flaxseed oil can become too hard and brittle, causing it to flake off (earning it the moniker "flakeseed oil"). That's why those who use flaxseed oil often combine it with another oil to make it more pliable.

Experiment with various oils or blends to see which works best for you.

Seasoning Methods

THE QUICK METHOD

Supplies
- · Oil of choice
- · Two clean lint-free cloths
- · Heat-resistant gloves (e.g. silicone)

Instructions
- · Apply a small amount of oil to the pan using a lint-free cloth. Then wipe off all the excess oil. This is important because too much oil can create a sticky film on the pan. If you have a cast-iron lid, treat it the same way.
- · Set the oven's temperature to a few degrees below the smoke point of the oil. Place the pan in the oven while it's getting to temperature and bake it for one hour. Turn off the oven and let the pan cool inside. You may need

to repeat this process two or three times before you feel the seasoning is sufficient. Since you won't be carbonizing the pan (heating past the smoke point of the oil), the pan may not turn dark black. That's all right. You can still use the pan with great results.

· Many sources recommend turning the pan upside down in the oven with a sheet pan underneath it to catch any oil that may drip off while baking. However, if you wipe the pan thoroughly, you shouldn't have excess oil on the pan.

THE PREHEATED METHOD

Supplies
· Piece of cardboard
· Heavy-duty welding gloves
· Oil of choice
· Two clean lint-free cloths

Instructions
· Skilletheads recommend preheating the pan before applying the oil. This is supposed to open the pores of the cast iron and prepare it for the seasoning process. The only issue is that you'll have to handle a hot pan.

· Start by placing the dry, stripped pan in the oven at about 250°F for fifteen minutes. Place a piece of cardboard on your counter to protect it. Then carefully remove the pan using heat-resistant gloves and place it on the cardboard. Typically, silicone gloves are recommended for handling cast iron, but this process is more prolonged, so heavy-duty welding gloves are recommended.

· Using a lint-free cloth, apply a small amount of oil to the cast iron. Wipe off any excess oil with a clean cloth and place the pan back in the oven. Raise the temperature to just a few degrees lower than the smoke point for the oil you're using and bake the pan for one hour. Then turn off the oven and allow the pan to cool inside.

· Repeat this process as needed.

THE FIRST MEAL

It's a good idea to pamper your freshly seasoned pan. For the first few meals, avoid acidic foods like wine and tomato sauce, which can destroy a thin layer of seasoning. Instead, feed your seasoning by deep-frying veggies or doughnuts. (See chap. 5, "Recipes," for ideas.) Stargazer also recommends using a little extra oil for the first six to ten meals.

Daily Conditioning

It's not difficult to keep a pan in tip-top shape.

CLEAN IT

Yes, you can use mild dish soap on a cast-iron pan! The soap our grandmothers used contained lye, which could strip a pan in minutes. The soap we use today is actually a mild detergent, and it's perfectly fine on cast iron. Just skip the scouring pad and use a gentle scrubber or silicone scraper instead.

KEEP IT DRY

After you wash your pan, dry it with a towel and then place it on the stove on low-medium heat until it's thoroughly dry.

SEASON IT

After drying the pan on the stove, perform a light seasoning. Add half a tablespoon of oil (or your favorite seasoning blend) to the pan and rub it into the cooking surface using paper towels and tongs to protect your fingers. Let the oil heat on the pan for two to five minutes. Get the temperature high enough to heat the oil but a bit lower than the smoke point. Turn off the heat and wipe off any oil residue. You can use the remaining oil to wipe the outside of the pan and handle if you'd like.

STORE IT

Place the cool, clean pan in a dry cabinet or hang it on the wall. If you store it on your counter, use a trivet to keep the pan from any moisture that could be on the counter.

Extreme Restoration

Once you get the hang of restoring cast-iron pans, you may want to try your hand at something a little more complicated.

UNUSUAL PIECES

Skillets and griddles are considered the easiest types of cast iron to restore, thanks to their easy-to-hang handle and flat surface area.

That's not the case with cornstick pans, gem pans, tea kettles, waffle irons, humidifiers, and similar pieces. Every Skillethead I spoke with had a run-in with an angry rust bucket. Here are some of my favorite quotes:

"Cornstick pans are as terrible as everyone says they are." —Ferris

"By far the hardest was an old number 8 unmarked humidifier (tea kettle, though, they are not). That thing fought me tooth and nail. Scrubbing the inside walls? NOPE, never again. 'Til the next time. Haha." —Cast Iron Kev

"The most difficult items to restore have yet to come, but are a very ornate matching set of candelabras that are currently in line to go into the electrolysis tanks!" —Orphaned Iron

"John Wright Co. (Lancaster, PA) made a dinosaur gem pan in the '80s or '90s. It is BY FAR the most difficult item to restore ever made. I wouldn't do another unless I HAD to." —Cast Iron Savannah

If you find yourself having to restore a particularly difficult item, you may need to reach for Easy-Off, Carbon-Off!, or Evapo-Rust, as described earlier in this chapter. However, there is no substitute for plain old elbow grease.

"Soap is by nature a de-greaser, a.k.a. a de-moisturizer, and while it won't strip away seasoning, it will dry out the surface of your pan. So only use it if you have to, and if you do, be sure to remoisturize the pan."

—Field Company's website

> "At the end of the day, people who want to start restoring or collecting cast-iron cookware are going to have to put in lots of time learning, trying, and failing before they're comfortable with it."
>
> —Ferris

BIG ITEMS

If you can't fit a cast-iron piece in your E-Tank, what do you do? Build a bigger E-Tank.

Although there are spray-on products available, like Easy-Off, it could take a dozen cans or more to cover a large item, making it cost-prohibitive. A Vinegar Bath is not a good option either because you still have to find a container to submerge it in, and you have to lift it up to check on it every thirty minutes. So a big E-Tank is still the best way to go.

One method is to build a wooden crate and line it with heavy-duty plastic. The plastic will then hold the water and the cast iron while the crate holds it all together. Another option is to use a large IBC tote, which is a big water container with metal reinforcement. Since the container is made to hold water, it's probably a safer setup.

Either way, though, you'll likely need a winch to help you hoist the item in and out of the water. Short of that, you could hang the item from a long board and get a friend or two to help you lift it into the container.

The actual method of electrolysis is the same as described earlier in this chapter. Just make sure you use enough anodes to cover the surface area.

Once the item is cleaned, it has to be re-seasoned. If the item is too big for your oven, you could suspend it over a sturdy tree limb over a small fire or a few propane burners. Use a handheld temperature gun to monitor the temperature and adjust the heat as necessary. After about an hour, the oil should be baked into the cast iron, both the side facing the burners and the opposite side as well. You can repeat the seasoning process as needed.

SYRUP KETTLES

The syrup kettle pictured here belonged to my husband's grandparents. They used it as a "hog scalder," to boil laundry, and even to water the cows. A much larger kettle was used for making cane syrup; that one ended up in a museum.

These old cast-iron kettles were intended as outdoor cookware, so they were seasoned just like skillets. It would be a huge undertaking to restore one this

This kettle is big enough to hold me and my son Gordon!

size, including building an E-Tank to hold it and a scaffold strong enough to hoist it in and out of the tank. That must be why I can't find anyone who restores these big kettles. Plus, I can't imagine ever making syrup in it again, so there's really no need to season it.

Instead, I'd like to turn this old hog scalder into a beautiful fountain one day. For that, I may simply scrub it and paint it a matte black.

If you have an old, rusty kettle that's too big to restore, consider repurposing it instead.

WOOD STOVES AND HEATERS

If you like vintage cookware, then you're sure to love antique wood stoves. But the process for restoring a wood stove can be a lot more complicated than a simple pan. To learn more, I spoke with Richard Richardson, founder of the Good Time Stove Company in Goshen, Massachusetts, where he's been restoring antique stoves since 1973.

Here in Florida, the most common wood stoves and heaters were from Atlanta Stove Works, the closest foundry. The stoves were simple but functional, and the heaters were essentially cast-iron fire boxes with two burners on top. Efficient, yes, but not fancy. The stoves and heaters Richardson showcases on his website—now, those are true works of art with interesting shapes, intricate designs, and bright metal trims in silver, brass, and copper.

I asked Richardson about this disparity, and he explained that the cast-iron foundries up north were founded long before the foundries in the southern states. That gave them time to develop more sophisticated cooking stoves. They also focused on artistry as much as efficiency, resulting in beautiful—yet functional—cast-iron stoves as well as heaters.

It's that same artistry and tangible, functional history that caused Richardson to start collecting stoves as a young man. At only twenty-four years old, he looked around and realized he had "too many stoves and not enough money to pay the bills." His solution? He restored the stoves he had and sold them. Then he found more stoves and repeated the process. It didn't take long for Richardson to realize he had discovered a niche market. "I fell in love with stoves and the industry, what we manufacture [in the United States], and the importance of them, and keeping them alive. . . . It's a lost industry."

Take a look at the online inventory at goodtimestove.com and you'll fall in love with antique stoves, too. But don't let their fancy looks fool you; these stoves and heaters are fully restored and ready to be used. They can even be converted to accommodate gas, propane, or electricity. Either way, Richardson is

1880s Wood Stove

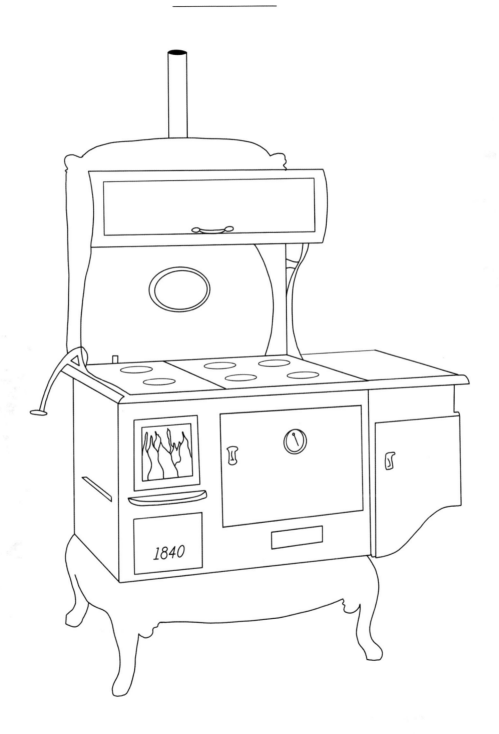

1880s Wood Stove, Detailed.

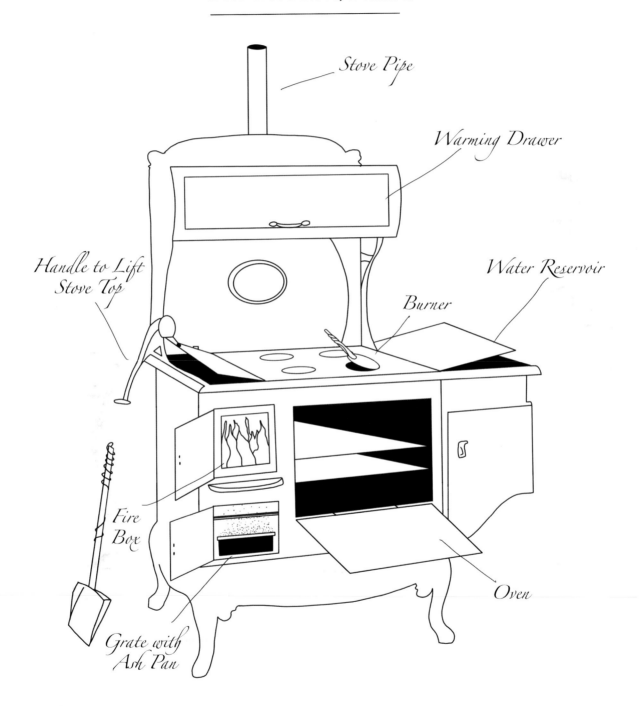

Stove Pipe

Warming Drawer

Handle to Lift
Stove Top

Water Reservoir

Burner

Fire
Box

Oven

Grate with
Ash Pan

Vintage stoves. *Courtesy Good Time Stove Company.*

convinced these old stoves are more efficient than modern appliances. He claims the manufacturers "reinvented the wheel and there was no reason"—and since he's been in business nearly fifty years, it seems his customers agree with him.

DIY INFO

If you have a simple wood heater or stove *without* windows, frames, or trim, you may be able to restore it at home. Here are Richardson's tips for at-home restoration:

1. Make sure the stove is not cracked or noticeably warped. If defects are noted, the piece should be used for decorative purposes only for safety reasons.

Atlanta Stove Company Stove

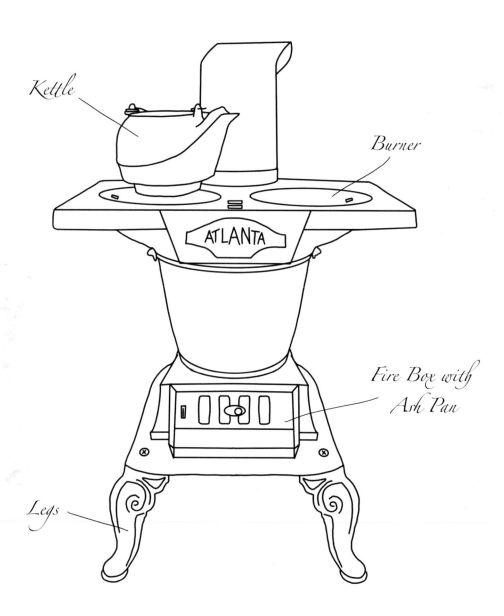

Kettle

Burner

ATLANTA

Fire Box with
Ash Pan

Legs

2. Remove the rust using a coarse rotary wire brush or an electric drill. (Richardson assured me this would not leave tool marks on the stove.)

3. If you'd like to minimize the amount of elbow grease needed, have it professionally sandblasted—but not with sand. Instead, have a professional use carborundum crystals, which are less coarse than regular sand and are often used to polish gravestones.

4. Polish the cooking surface with stove polish such as Rutland's Stove Polish in black.

5. Paint the exterior parts of the stove with a high-temperature paint. Richardson recommends Rutland Hi-Temp Paint in flat black.

While you can certainly polish the entire stove, Richardson recommends painting it instead because paint covers well and holds up better than polish. When you're done, buff the paint to make it shine.

If you'd prefer to have a professional restore your wood stove or heater, contact Good Times Stove Company for an estimate.

WHY IS YOUR WOOD STOVE BLACK?

Cast-iron cookware is black due to the baked-on seasoning. However, cast-iron stoves aren't seasoned. So why are they black? Historically, manufacturers polished stoves and heaters black until high-temperature paint became available. If you're restoring a wood stove, choose the product that fits your needs and aesthetics.

"Shop until you find a good stove. Don't buy paperweights."

—Richard Richardson, Good Time Stove Company

My favorite pans. *Photo by Courtney Wahl.*

Meet the Skilletheads

It may surprise you to learn that there is an entire community of people who enjoy nothing more than removing rust off an old skillet, but it's true.

I wrote my first book, *Modern Cast Iron*, from my perspective as a home cook, busy mom, and cast-iron connoisseur. As I shared my journey, cooking tips, and recipes online, I connected with other cast iron fans from around the world— and what a group! Made up of home cooks like myself, retired grandfathers, busy dads, and young people just discovering the value of cast iron, this community embraces anyone who shows an interest in the black pan of our forefathers.

But within this community is something extra special: a thriving subset of cast iron restorers, collectors, and sellers. These folks spend their spare time hunting down rare and quality pieces, going to great lengths to restore them to their former luster, and then selling them to other cast-iron fans. When not working on cast iron, they're often collecting old or new pieces for their own use and to pass down to their kids. They start the websites, forums, and online groups to help folks like us understand how to use and care for cast iron. These are the people known as Skilletheads.

In this section, you'll get to meet the Skilletheads I've been working with. These folks have taken the time to share their personal stories, restoration methods, preferred seasoning oils, favorite cast-iron brands, and more. Since every Skillethead has something different to offer, I've included the following symbols on each page for your reference:

 Provides information or resources on cast iron

$ Sells restored cast iron, ancillary products, or food

Restores cast iron for a fee

Shares recipes

Whether you're looking for more information, a little inspiration, or to hire someone to restore your late grandmother's pan, these Skilletheads are ready to help.

Cast Iron Kev $

Etsy etsy.com/shop/castironkev
YouTube youtube.com/user/kekermahoney
Facebook, Instagram, and TikTok CastIronKev
Location New Jersey
"NJ. Vintage cookware, vintage everything!
Keeping usable history alive one piece at a time."

vintage.iron.cookware

If you like all things vintage, you're sure to like Kevin Fogarty, a.k.a. Cast Iron Kev. His posts and videos are proof that cast-iron restoration can be a fun—and dare I say exciting?—hobby. If you like the thrill of the flea market hunt, check out his videos where he picks through piles of cast-iron pans to find the ones worthy of a second chance. He also restores the occasional enamelware or Pyrex piece using a food-grade Lye Bath.

Kevin has been restoring cast-iron cookware for several years. What once was an interesting hobby is now a part-time job, and he gives his wife all the credit. Kevin told me that when they were dating, she objected to his scratched Teflon and dented pots and pans. They did some research and discovered that cast iron was a great alternative. "I dug deeper and realized that older pieces were lighter, smoother, and can be found needing work at a discount! We're into DIY, so I purchased a three-piece set of rusty Wagners and tried it out. They weren't pretty, but they did the job. Once our family and friends saw us using them, they wanted sets themselves. And that's totally my excuse for jumping into this!"

And jump he did. At one point Kevin had 154 "keepers" in his cast-iron collection. "I knew it was on the way to out of control! So I only keep what we use, which is about 20 various pieces." That's why he prefers restoration over collection. "Bringing back a skillet that is sure to end up in a landfill or melted down is a great feeling. And you'll give me money when I'm done? Sign me up!"

For those pieces Kevin can't restore—the chipped, cracked, or warped ones—he gives them new life by turning them into cast-iron spatulas. These unique utensils are great for grilling.

Check out Cast Iron Kev's Etsy page for his current inventory of restored cast iron, other vintage cookware, and cast-iron spatulas, and follow "the hunt" for more cast iron on his YouTube channel. If you live in New Jersey, you can also reach out to Kevin to have your cast iron restored for a fee.

Restoration method

Food-grade Lye Bath followed by a Vinegar Bath. If the Vinegar Bath is insufficient, he uses Evapo-Rust. He plans to set up his first E-Tank in late 2022.

Kevin's son Deran is the cutest Skillethead on the block. *Courtesy Cast Iron Kev.*

Handmade cast-iron spatulas. *Courtesy Cast Iron Kev.*

Seasoning

Cast Iron Kev uses flaxseed oil to lightly season the pans he sells in his Etsy shop. Buyers are then able to perform additional seasonings at home using their preferred oils.

At home, Cast Iron Kev reaches for Crisco or even a spray like Pam; he calls these oils "old-fashioned, simple, strong, and inexpensive." However, he does recommend seasoning brands like BuzzyWaxx, Crisbee, and Easy Beezy. "They're just not cost effective for myself. (Too many pieces per month!)"

> "I am stubborn, and I will not be defeated by crust or rust!"
>
> —Cast Iron Kev

Favorite brand

Wagner. "But I can find beauty from any foundry."

Regarding new cast iron . . .
Cast Iron Kev is a diehard restorer—he proudly states that he's never bought a new piece of cast-iron cookware.

Advice for new restorers
Don't trust Google for everything. Join a free group online and ask questions. And if you really like the idea of restoration, you may want to consider it for your next part-time job. As Cast Iron Kev says, "It doesn't feel like work if you love what you do!"

Advice for new collectors
Put yourself out there and be persistent. "Nobody drives farther, gets there earlier, or walks faster than me!"

Cast Iron Savannah

Website CastIronSavannah.com
Facebook, Instagram, and TikTok castironsavannah
Location Savannah, Georgia
"Everything tastes better in cast iron."

Now located in Savannah, Georgia, Ken Margraff and his wife, Jes, are originally from Pennsylvania, home of the legendary Griswold. As they state on their website, "Both of our fondest memories are the times spent in our family's kitchens learning how to cook and bake."

However, their journey with cast iron didn't start until recently, when Jes gave Ken a modern Lodge 8" skillet for Hanukkah/Christmas. Ken told me his story: "I became obsessed with using it and shortly after found a rusty Lodge cactus gem pan at Goodwill. I started to research how to restore it and got hooked. From there it was down the rabbit hole I went. Once I started getting good at it, friends and family wanted me to restore items. By that point, I was buying every piece of iron I could find. Somehow my business was born out of that chaos."

Before-and-after picture. *Courtesy Cast Iron Savannah.*

An avid collector, Ken gets a kick out of the "thrill of the kill," or hunting down valuable (or just plain fun) cast-iron pieces to restore and sell. Still, he claims, "My favorite part is seeing the before and after from restoration."

And that's a good thing, because Ken does a lot of restoration. If you're looking for something specific, simply email the duo through their website, and they'll search through their inventory of over one thousand pieces! And if you'd like to have a special pan restored for a fee, they can do that, too.

Ken's big collection of miniatures. *Courtesy Cast Iron Savannah.*

What's next for Cast Iron Savannah? Ken said they'd like to build a workshop in the backyard. That would allow them to re-season pans outside and give them more room to restore larger items like kettles.

Restoration method
Ken uses a combination of Lye Baths, Evapo-Rust, and E-Tanks to restore cast iron.

Seasoning
Ken applies at least two coats of BuzzyWaxx seasoning.

Favorite brand

"Lodge. They have been around for over 125 years, never closing. They survived multiple world wars, pandemics, depressions, and everything else thrown their way. To me, they are the perfect example of the American spirit."

And Ken's favorite piece in his collection? "That's like picking a favorite child! My favorite USER is my large egg logo Lodge 2 quart casserole pan. My favorite collector items are my Lodge mini collection."

Regarding new cast iron . . .

Ken is fond of Stargazer in addition to modern Lodge.

Advice for new restorers

"AVOID YouTube and Reddit," which are rife with bad advice.

"Don't do it alone. Join a community to learn the correct ways to do things. On top of knowledge, you will find some of the most amazing friends you could ever ask for."

Advice for new collectors

If you aren't sure about a piece, "snap a picture and post it to the Cast Iron Community Facebook page while hunting. Don't put it down or leave until someone responds!"

Cast Iron Steve

Website CastIronSteve57.wixsite.com/website
YouTube youtube.com/channel
/UCbdmoBxTDf-MUXKLtcGQajw
Facebook CastIronSteve
Instagram cast_iron_steve
Location Kaysville, Utah
"I'm just a young guy that likes to cook in cast iron like the old-timers but with a modern twist."

Steven Brewer, a.k.a. Cast Iron Steve, may only be in his twenties, but he's been cooking with cast iron for over a decade. At only twelve years old, he went on his first Boy Scout camp out and became intrigued with the cast-iron skillets and Dutch ovens the Scout leaders used. He marveled at how good the food tasted and then wondered why the leaders oiled down the heavy cookware after they used it. Over the next few years, Steve's interest grew until he purchased his first Dutch oven—a 12" Camp Chef—at the ripe old age of sixteen. He still uses that Dutch oven whenever he goes camping with his family.

I asked Steven if he puts his Dutch oven right on the coals or if he uses a fancy tripod to suspend it over a fire. He said he usually just places it on the coals, but he has used a tripod several times and each time he thinks, "Why don't I use this more? Just out of the fact that it looks like something you'd see in an old Western movie. It just looks cool." He said the food tastes the same either way, though.

When it comes to home cooking, Steven said he uses his cast iron "every single day," whether it's on the stove, in the oven, or on his pellet grill. He even uses the grill to season his pans. Steven hasn't jumped into cast-iron restoration—"yet!"—but that might be on the horizon.

Cast Iron Steve has quite the following on social media, and it's all for the love of cast iron, grilling, and good food. Follow along for inspiration on what to make for dinner, how to season your pan, and other grilling tips and ideas, as well as a discount code for BuzzyWaxx seasoning.

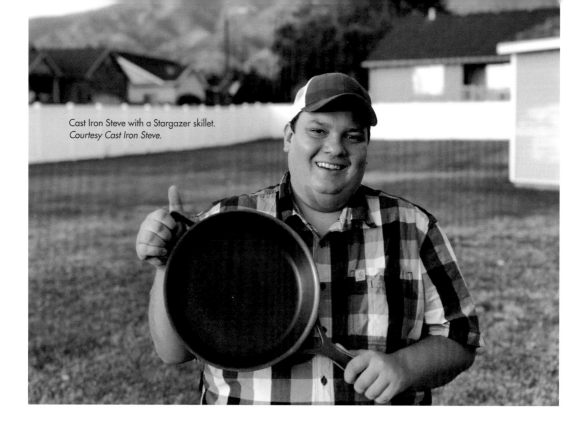

Cast Iron Steve with a Stargazer skillet.
Courtesy Cast Iron Steve.

Seasoning

Cast Iron Steve seasons all his pans with BuzzyWaxx. He states it creates a better nonstick surface than other oils. He's also diligent about performing a light seasoning after every use and then a deep seasoning on the pellet grill about once a month. The grill allows him to use a higher temperature while keeping the fumes outside.

Favorite brand

Cast Iron Steve uses a lot of brands, including Lodge, Camp Chef, and some vintage cast-iron pieces, but his favorite skillets are by Stargazer. "I stand behind their product. It cooks great and holds temperature." His favorite Dutch oven is still the Camp Chef 12" that he bought when he was sixteen years old.

Performing a deep seasoning on the pellet grill. *Courtesy Cast Iron Steve.*

Advice for new cast-iron users

"Don't overcomplicate it. Cast-iron cooking is super easy. The only difference between cast iron and other pans out there is you have to season them. If you do this after every use, it only takes a matter of minutes."

"A Skillethead is someone who has more cast iron than they need but they want more. My wife says I have a problem."

—Cast Iron Steve

Chef Lance McGinnis $ 🍴

Website old-skool-by-gritz-n-greenz.square.site
Facebook and Instagram gritzngreenz
Location Tallahassee, Florida
"Southern Food with Attitude"

When I heard that Chef Lance McGinnis of the Gritz N Greenz Old Skool food truck loves to cook with cast iron, I had to know how he managed it. Did he cook in multiple pans at one time? Did he clean the pan between each use or simply wipe it down and do a thorough cleaning at the end of the day?

It turns out Chef Lance doesn't use cast iron in his food truck—for those cook-to-order meals, anyway—due to those logistics issues. However, he does use it for special events and when preparing certain foods ahead of time, especially cornbread.

I met up with Chef Lance at The Prepared Table Kitchen Shop in Tallahassee, where he was giving a cooking demonstration. The menu was dry-aged steak with mushrooms and onions, bacon-wrapped asparagus, golden potatoes baked with bacon and cheese, and a dessert of berries with homemade whipped cream. Everything except the dessert was cooked in cast iron (Smithey Ironware, to be exact), and it was all amazing! Thankfully, Chef Lance agreed to share his recipes with us, so look for those in the "Recipes" chapter.

While the steak was resting and we were waiting on the guests to arrive, Chef Lance spoke with me for a few minutes about his appreciation for cast iron, especially the "durability and the evenness of the cooking. So from edge to edge, you're not getting a cold spot or a hot spot." He then shared with me his two culinary passions: southern comfort food and cooking for metabolic health. When it comes to cast-iron cookware, Chef Lance gets the best of both worlds.

"There's nothing like bacon in cast iron."

—Chef Lance McGinnis

He can cook all the yummy foods he craves (like steak and potatoes) but doesn't have to worry about eating chemicals or inhaling fumes, which is possible with chemical nonstick pans. As he explained, cast iron is "a natural element from the earth. It's just iron forged. And that's the beauty of it. We talk about holistic health. It's kind of like a holistic part of cooking."

Chef Lance and I at a cooking demonstration. *Author's photo.*

Chef Lance wrapping asparagus with bacon. *Author's photo.*

If you're in the Tallahassee area, check out the Gritz N Greenz Facebook page to find out where the Old Skool food truck is located that day. You're sure to love his southern menu, especially the Shrimp N Gritz or the Chicken N Waffles. And if you're looking for a special date night, sign up for one of Chef Lance's cooking demonstrations at tptkitchenshop.com.

Seasoning
Chef Lance seasons with coconut oil, avocado oil, or even bacon fat if he plans to use the pan again soon. However, he avoids vegetable and seed oils because they're more processed.

Favorite brand
Smithey Ironware. Chef Lance has become especially fond of their Farmhouse Skillet, which is actually made of carbon steel.

Cooking tip
For perfect cornbread, heat the pan with a little oil until it's "hot and sizzling." Then pour in your batter and bake as usual.

Ferris   FERRIS

Website FerrisCastIron.com
Facebook, Instagram, Twitter, and Pinterest ferriscastiron
Location Alexandria, Virginia
"We restore heirloom-quality cast-iron cookware and teach home cooks how to use and care for it."

Nick Turner is the man behind Ferris, a cast-iron cookware restoration business out of Alexandria, Virginia. When I asked Nick why he started restoring cast iron, he talked about growing up in the 1980s and 1990s, "when environmentalism and sustainability went mainstream." Nick shared his story:

Rusted vintage Lodge Dutch oven. *Courtesy Ferris.*

Lodge Dutch oven after restoration. *Courtesy Ferris.*

"All the while, my mother was teaching me that the companies you support with your spending should reflect your values. So, I grew up to be the kind of person who cares about the environment and who thinks carefully about where he spends his money.

"When I learned about the dangers of nonstick cookware, all of that came together to lead me to cast iron. Cast iron is everything nonstick cookware isn't: nontoxic, long-lasting, and versatile. On top of all that, some brands' foundries are located just a few hours' drive from my home, allowing me to buy from companies in my part of the country."

"What other cookware or collectible is safe, locally made, versatile, and also so durable that you can use it daily without destroying its value?"

—Ferris

When Nick threw out his nonstick cookware, he replaced it with several pieces from Lodge. "Eventually, someone wasn't careful enough and got one of the skillets rusty (thanks, kids!), and figuring out what to do about that took me down an internet rabbit hole of methods, myths, science, and history. It wasn't long before I was buying vintage pieces anywhere I could find them, restoring them, and selling them or giving them to friends and family as gifts."

Like most restorers, what started as a hobby soon became a business. But the satisfaction of restoring a piece of history enables Nick to enjoy each piece he works on. "There's something deeply satisfying about resurrecting a rusted, crusty piece of cast iron. There's a lot of artistry in even the simplest pieces, and it feels really good to reveal that beauty again after it was hidden by time and wear. Also, you can't beat the feeling of rubbing oil into a slightly warmed skillet with your bare hands."

Nick loves to share his passion for cast iron online and in person. If you're in the Alexandria area, check out the cast iron cooking classes and other services he offers in addition to restoration. Ferris usually serves local customers, but if you'd like to ship your pan to Ferris for restoration, contact Nick for an estimate and directions. And for those restored-and-ready pans, check out the Ferris website and eBay listings.

Restoration method

"We use short, timed soaks in a solution of organic white vinegar and water to remove rust when it covers about a third or more of a piece of cookware. If there are just a few spots, we'll do spot treatments with the same solution. To get rid of burned-on food and old seasoning, we use extended soaks in lye and water."

Seasoning

Organic, refined flaxseed oil baked between 450°F and 500°F.

Favorite brands

"I'm amazed by Butter Pat's ability to produce smooth cookware that doesn't require machining—they don't grind or mill it smooth afterward, they cast it smooth! They're my favorite 'modern vintage' brand at the moment. My favorite vintage brand is Birmingham Stove & Range."

Regarding new cast iron . . .

"In the last year, I've bought a Field Company #10 lid, a Butter Pat Heather skillet and glass lid, and a Smithey Ironware #12 griddle. I'd love to pick up a Dutch oven from Borough Furnace and a sauce pot (or two) from FINEX."

Advice for new restorers

"I suggest buying a low-price modern piece of cookware to practice on before you try on an heirloom-quality vintage piece."

Advice for new collectors

"Don't let other collectors decide for you which brand or manufacturer is valuable to you. The heart wants what the heart wants, and that's just as true for cast-iron collecting as it is in love."

Fired Cast Iron

Website FiredCastIron.com
Facebook EnameledCastIronCooking.com
Location Red House, West Virginia
"Fired Cast Iron - Rustic Cooking - Simple Living"

Although Jeff and Dee Holbrook have cooked with seasoned and enameled cast iron for years, it wasn't until they retired in 2020 that they started their YouTube channel, FiredCastIron.com. Jeff says, "Now we simply have time to indulge our passion for cooking with cast iron and to tell more people than just our close friends."

The couple use their glass-top electric range and stove for most meals, but Dee also enjoys cooking in their indoor 1978 Vermont Castings stove and outside on

Jeff and Dee Holbrook of Fired Cast Iron.

their 1923 Atlanta Stove Works stove. By sharing their videos online, they hope to preserve a record of Dee's rustic cooking style and to share her methods with others.

The Holbrooks are quick to share that Dee has a disability and Jeff has mobility issues, but they don't let that stop them from using heavy cast iron. Instead, they've learned to cook together—and they've found that they enjoy working as a pair.

But why cast iron? "Practical use proves that the heat retention of cast iron lends itself well to the rustic, simple style of cooking we like to do. We don't own the latest fancy trends in specialty cooking machines. We love to take the time to cook together, and the quick and easy machines of cooking that people use to save time take away from the experience." The quality of the food and of the cooking experience itself—this seems to be a common thread among cast-iron fans.

What's next for Jeff and Dee? They recently purchased an RV, so "more seasoned cast iron videos will be forthcoming from our travels." Follow along at FiredCastIron.com. And if you enjoy cooking with enamelware, join their Facebook community at EnameledCastIronCooking.com.

Favorite brands

For seasoned cast iron, the Holbrooks stick with Lodge. They can fit a Lodge 10" deep skillet with matching lid inside their indoor stove, which is a perfect setup for baked beans, potatoes, and pot roast. They use a 12" Lodge skillet on the burner of their outdoor stove to make dishes like roasted veggies.

For enamelware, Jeff and Dee typically buy only new pieces to avoid the potential for lead that manufacturers used years ago. Their preferred brands of enamelware are from chefs Mario Batali, Rachael Ray, and Wolfgang Puck. They also like KitchenAid and Cuisinart, stating that all of these provide "consistent quality."

Beautiful collection of cast-iron enamelware. *Courtesy Fired Cast Iron.*

Advice for new collectors

The Holbrooks admit to having a good collection of seasoned and enameled cast iron, but they are not collectors. "Collecting is not in our minds. It is all about the usefulness of the items." Jeff and Dee have simply chosen specific seasoned and enameled pieces over the years that help them cook the meals they enjoy, and they encourage others to do the same.

If you are looking for enamelware, the Holbrooks recommend shopping at TJ Maxx, HomeGoods, and Marshalls for good-quality pieces at low prices. You'll have to make one concession, though: you'll end up with a mix of colors. If you want them all to match, you'll have to pay full price.

Kent Rollins **$**

Website KentRollins.com
YouTube youtube.com/cowboykentrollins
Facebook and Instagram cowboykentrollins
Location Red River Ranch near Hollis, Oklahoma
"Kent Rollins is from a lost period in time and a dying state of mind, when life was simple and character was king."

Kent Rollins is, among other things, a chuck wagon cook, caterer, cookbook author, storyteller, TV sensation, and YouTube celebrity. At the beginning of his career, Kent helped his uncle guide and feed hunters in the Gila Wilderness of New Mexico. Later, he bought his own chuck wagon and began cooking for cowboys working on ranches. Kent and his wife, Shannon, now travel the country sharing recipes and a simpler way of life, and Kent continues to cook for cowboys on ranches during spring and fall gatherings.

If you've seen any of Kent's videos, you've probably seen him cook with cast iron. I asked him recently if he grew up in a house with cast-iron cookware or if he learned to use it later in life. He said, "Mama always had a cast-iron skillet on the stove. It was used to create many a meal growing up. After school we'd run home and see what leftovers she had waiting on the stove from Dad's lunch—typically they were still warm in the skillet."

Kent Rollins with his chuck wagon. *Courtesy Kent Rollins.*

While Kent and professional restorers have a love of cast iron in common, they do not agree fully on the methods of restoration. In his videos, Kent has promoted the use of a self-cleaning oven and even tossing the pan in a fire when all else fails.

"Can't get full on fancy."
—Kent Rollins

I asked Kent why he never uses an E-Tank, the preferred tool of many restorers. His answer: "We don't use electrolysis because there are other ways of restoring cast iron that don't require a lot of equipment. We keep things pretty simple."

That makes sense. When Kent hitches up the chuck wagon and heads out to a ranch for a few days, he can't very well take an E-Tank or a Lye Bath with him. But he does have ready access to salt, vinegar, scrub brushes, and even fire when necessary—the same tools our grandparents had at their disposal.

Cooking in Bertha, Kent Rollins's famous outdoor stove. *Courtesy Kent Rollins.*

Kent may have a "simple" cowboy lifestyle, but it sure is interesting to those of us who live in a fast-paced world. Follow Kent's adventures and check out his recipes and cast-iron tips on his YouTube channel. You can also purchase Kent's cookbooks and his own line of spices, sauces, and coffee through his website.

Restoration method

In his video "Safely Remove Rust from Cast Iron," Kent demonstrates several methods of removing rust, including using vinegar and baking soda and even a wire drill. He also recommends using the self-cleaning cycle in the oven or placing the pan in an open fire as a last resort. However, he specifically states these last two methods are not recommended for antiques or thin pieces because the heat can crack the pan or increase warping.

Seasoning

Kent likes to use three base coats of flaxseed oil then a layer of avocado oil, baking each layer of seasoning at 300°F for forty minutes.

Favorite brands

Wagner, Griswold, Wardway (by Wagner), Field, Stargazer, and FINEX.

Advice for new cast-iron users

Kent recommends a 12" skillet and a 12" Dutch oven, either of which can be used as a substitute for a 9 × 13 pan. (Kent measures Dutch ovens like skillets, by the diameter of the opening.) At these sizes, they'll easily hold casseroles, biscuits, cakes, and pies, and the Dutch oven will be deep enough to bake breads as well as stews.

Orphaned Iron

Etsy etsy.com/shop/OrphanedIron
Facebook, Instagram, and TikTok OrphanedIron
Location New Paris, Ohio
"A company dedicated to restoring and offering antique Cast-Iron Cookware."

There's a reason why Skilletheads restore cast-iron cookware and not some other cookware or collectible item: they appreciate the history, the intimate memories that are embedded in each pan. Memories of Grandma's fried chicken Sunday dinners and Great-Grandma's famous biscuits. Of children laughing and families coming together around the table. For Skilletheads, each piece of old cast iron is a treasure reminding us of a simpler life and the value of family.

When I asked Matt Bright of Orphaned Iron about his unique business name, I wasn't surprised when he waxed poetic:

"Every antique pan that you see at a yard sale, flea market, auction, or antique mall was once part of a family. The mass majority of them were kept in the family for decades and used for every meal. Cast-iron pans are still a staple in descriptive memories of meals prepared by grandparents. Somewhere along the line, those pans were discarded and left without a family. Restoring antique pans and finding a new home for them is something I truly cherish. I'm in the business of finding new homes for Orphaned Iron."

For Matt, this rescue mission is personal. He learned all about using and caring for cast iron from his father, who learned it from his grandparents. "After my father passed, I focused on restoring cast iron as a way to keep his memory alive."

"The passion is what matters, not the size or value of a collection."

—Orphaned Iron

Matt started Orphaned Iron after purchasing a collection of two hundred cast-iron pieces from Facebook Marketplace. Now, he restores and sells cast iron as a full-time job through Etsy (and has a private collection of around four hundred pieces!).

If you'd like to watch Matt's tips and tricks for restoring and caring for cast iron, check out the

A part of Orphaned Iron's collection.
Courtesy Orphaned Iron.

Cooking with a restored Griswold loaf pan. *Courtesy Orphaned Iron.*

Orphaned Iron TikTok channel. And if you decide you want a pro to tackle your cast-iron pan, Orphaned Iron is ready to help.

Restoration method

Matt uses only an E-Tank to restore cast iron. He said it removes old seasoning, paint, carbon, and crud as well as rust.

Seasoning

Matt prefers to season his cast iron with Crisco shortening. "I've tried multiple oils over the years and Crisco seems to hold up the best."

Favorite brand

The Favorite. "These items were produced from 1882 to 1902 in the Ohio Penitentiary by inmates in Columbus, Ohio. Few and far between, but a truly interesting history behind them."

Regarding new cast iron . . .

Matt never buys it. He prefers rescuing old cast-iron pieces instead.

Advice for new restorers

"Do some test runs on newer Lodge pans or Asian-made pans before attempting high-end collectibles."

Advice for new collectors

Buy only what you enjoy; don't worry about what is most popular or valuable.

Skillethead John

Location Munford, Tennessee
"Cast iron collecting, restoration, and some southern cooking from da dirty south."

The first time I saw the term *skillethead* was when I came across Skillethead John, a.k.a. John Wolfe. He used to restore and sell cast iron through his Instagram account, but he recently got off social media. He still restores cast iron for himself every once in a while, but he no longer sells restored pieces.

I miss seeing John's posts of homemade dinners, his garden, and his family. My favorite was the time he shared a picture of his well-seasoned cast-iron pans hanging from ropes in the backyard like large wind chimes. Apparently, his goal was to use the clanging of the cast iron to keep the birds out of his fruit trees! I commented that shiny aluminum pie plates usually work well, and he responded with a laugh: "Why didn't I think of that? Well, at least I got to work with my cast iron in the garden this weekend."

Collection of cast iron, some rusty, some restored. *Courtesy Skillethead John.*

Restored and ready to use.
Courtesy Skillethead John.

To me, John's focus on family and friends (and his gentle humor) makes him the quintessential Skillethead. In his own words, cast-iron restoration is "a simple hobby that includes your family and friends. Good meals means happy people."

A while back, I asked John how he got started restoring cast iron, and he said it began with a financial need. "We were going through some rough financial times, and I wanted to cook more homemade meals. I saw two newer Lodge pans at my local grocery store and bought them. Started looking up how to clean and care for them and found some good Facebook groups that taught me correct methods of cleaning. I was hooked on cast iron then."

While you can't follow Skillethead John online anymore, I wanted to share his story and his restoration methods and tips with you. After all, preserving these methods is an important part of preserving cast iron.

Restoration method
Lye Bath followed by an E-Tank. "I then wash it with a stainless steel scrubby in cold water until everything is off the pan. I then do a cold water rinse with an SOS pad and finally with a Magic Eraser. The cold water helps to stop any flash rust if you live in a humid environment."

Seasoning
John makes his own seasoning blend of beeswax, canola oil, grapeseed oil, and extra virgin olive oil.

Favorite brands
Vintage all the way—Wagner, Wapak, Griswold, and Vollrath. His favorite skillet is a #2 Favorite Piqua Ware pan.

Regarding new cast iron . . .
John still likes his Lodge pans.

Advice for new collectors and restorers
"Character counts in this hobby. Treat people fairly."

What's Up Homer Skillet? $

Website WhatsUpHomerSkillet.com
Facebook and Instagram WhatsUpHomerSkillet
Location Athens, Georgia
"Collecting, restoring and sharing the love of cast iron cookware. Skillets, griddles, and Dutch ovens available. Restored and ready!"

Like most Skilletheads, David Ragsdale of What's Up Homer Skillet? has fond memories of family meals cooked in big cast-iron pans—a #8 Favorite, a #8 three-notch Lodge, and a #8 BSR, to be exact. Of course, as a child, he didn't understand the history, quality, and value of the skillets his parents used, but his appreciation grew after a friend gave him a #9 Griswold in college—a pan he uses to this day.

"Like Kudzu, cast iron can become an invasive species. When I'm in full restoration mode, I tend to take up the bulk of our kitchen with a cleaning station, a seasoning station, and a cooling station."

—What's Up Homer Skillet?

Over the past few years, cast iron has become increasingly important for David, from a passing interest to a part-time restoration job. The name What's Up Homer Skillet? is in honor of David's miniature poodle, Homer, with a nod to some of his favorite 1990s slang. Part family, part humor, and all business, What's Up Homer Skillet? sells "restored-and-ready" cast iron through social media and a website and locally at the Athens Antique and Vintage in Georgia.

Even if you haven't met David in person, though, you may have seen him in the viral video "Why People Love Cast Iron Pans (And Why I'm on the Fence)," by journalist Adam Ragusea. In the video, David gives us a glimpse at his E-Tank and shows some work in progress as well as a few collectible pieces he's restored. Ragusea might not be convinced that cast iron is the way to go, but we sure are!

If have a pan you'd like restored, reach out to David through his website or social media. He's a teacher by day and restorer by night, so availability is limited.

The What's Up Homer Skillet? booth at the Athens Antique and Vintage. *Courtesy What's Up Homer Skillet?*

Restoration method

After a Lye Bath, David cleans the piece thoroughly with Bar Keepers Friend and then places it in an E-Tank. If there's only surface rust, he may use a Vinegar Bath instead of the E-Tank. "I've generally stayed away from Easy-Off, although I know many people have great experience with it."

Seasoning

"I've tried traditional Crisco, spray-on Crisco, vegetable oil, avocado oil, flaxseed oil, the Crisbee Puck, and BuzzyWaxx. At this point, BuzzyWaxx is my go-to product." To apply his seasoning, David prefers using a Scott Shop Towel.

Favorite brands
Favorite and Griswold.

Regarding new cast iron . . .
David has used and enjoyed several modern cast-iron brands, including
Marquette Castings. Still, he admits, "my passion is the collection of and
restoration of vintage pieces."

Advice for new restorers
"My recommendation is to review as many resources as possible, experiment
with what you learn, and chart your own path. I mention on my website, there's
so much contradictory information, you just have to decide what works best for
you and for your collection."

Advice for new collectors
"Read, read, read. Join Facebook groups. Go to the library and research. Like any
hobby, this can be very expensive, so the more you educate yourself, the better
your purchases will be and the more fun you'll have."

Authentic 1800s hearth at the Tallahassee Museum. *Author's photo.*

Recipes

The following recipes were contributed by the Skilletheads I interviewed for this book, from manufacturers to restorers and foodies. Some recipes feature specific cast-iron brands or products, but you're welcome to use whatever cast iron you have on hand. For those recipes that utilize specialty grills, you can use a regular grill or your oven instead.

Cooking is all about experimenting and finding what makes us happy. I trust this collection of recipes will inspire you to try something new in your cast iron today.

Breakfast

Sunday Frittata

What's Up Homer Skillet?

David Ragsdale shared this special frittata recipe, based on his
father's classic Spanish omelet.

Directions

1. Preheat the oven to 400°F.

2. Crack eggs into a blender. Add half the salt and pepper and
   blend for 30 seconds on high. Let the eggs rest for 5 minutes.
   Then add milk and pulse 3–4 times. Set aside.

3. Heat 1 tablespoon of olive oil in the skillet and add the onion.
   Cook until translucent and tender.

4. While the onion is cooking, steam the potatoes in a Ziploc
   Zip 'n Steam bag for 3–4 minutes or until tender. Add the
   potatoes to the onions in the skillet. Drop in the remaining
   salt and the red pepper flakes, and stir with a wooden spoon
   until the potatoes begin to brown, roughly 5 minutes. Then
   stir in the greens, thyme, and garlic. Add optional ingredients
   to the skillet and stir the mixture for 2–3 minutes.

5. If you aren't confident in the nonstick ability of your skillet,
   transfer the ingredients to a large bowl and ensure the skillet
   and its sides are well lubricated with olive oil or the cooking
   spray of your choice. Return the ingredients to the skillet.
   Flatten with a wooden spoon into a uniform layer. Pour the
   egg mixture into the skillet over the vegetables. Make sure all

Equipment

blender

large cast-iron skillet

Ziploc Zip 'n Steam bag

Basic ingredients

8–10 large eggs

1 tsp. salt (ideally pickling
or sea salt), divided

½ tsp. ground pepper

½ cup whole milk

2 Tbsp. olive oil, divided

1 medium Vidalia onion,
diced finely

1 large Yukon gold potato,
diced into half-inch chunks

pinch of crushed red
pepper flakes (optional)

2 cups greens of your choice
(kale, collard, or spinach)

2 tsp. fresh thyme leaves

2 cloves garlic, minced

1 cup shredded cheese, such
as Colby Jack, cheddar, or
Havarti (optional)

Optional ingredients

diced bell peppers

sautéed mushrooms

steamed radishes

heirloom tomato tossed
with thyme, oregano,
basil, and olive oil with a
dash of salt and pepper

ingredients are entirely covered by the eggs. Sprinkle the top with cheese.

6. Move the skillet from the stove top to the middle rack of the oven and bake the frittata until the eggs are set, roughly 12–15 minutes. When a toothpick comes out clean, turn on the broiler for about 90 seconds to bring about a nice, crispy top. Once the top is browned, pull from the oven.

7. Let the skillet cool for at least 10 minutes. Using oven mitts, flip the skillet over a serving plate and gently tap the bottom to release the frittata, then slice and serve.

Courtesy Lodge Cast Iron.

Sausage-Stuffed Pancake Sticks

Lodge Cast Iron

Put those cornstick pans to good use with this fun breakfast idea that will satisfy eaters of all ages.

Directions

1. Preheat the oven to 400°F. Place two Lodge Cornstick Pans in the oven while oven is preheating.

2. Cook sausage links in a Lodge Cast-Iron Skillet according to package instructions. Set aside.

3. In a large mixing bowl, combine flour, sugar, baking powder, salt, buttermilk, egg, and 2 tablespoons of melted butter.

4. Carefully remove pans from the oven and divide the remaining 2 tablespoons of melted butter among the 12 cornstick wells. Next, place 2 tablespoons of pancake batter into each well and then place one sausage link in the batter. Pour another 2 tablespoons over each sausage link.

5. Bake for 15–18 minutes or until a toothpick inserted into a pancake stick comes out clean. Serve with your favorite syrup or sauce.

Equipment

2 Lodge Cornstick Pans

small Lodge Cast-Iron Skillet

Ingredients

12 sausage links

1 cup all-purpose flour

4 Tbsp. sugar

1½ tsp. baking powder

½ tsp. kosher salt

¾ cup buttermilk

1 egg

4 Tbsp. melted butter, divided

syrup or sauce for serving

Ricotta Dutch Baby Pancake

Smithey Ironware

It's a good thing this recipe makes two Dutch babies. Filled with ricotta cheese and topped with strawberries, one just isn't enough. The recipe calls for two No. 6 skillets, which are 11.25" in diameter. If you need to, substitute with pans of similar dimensions or make one small Dutch baby and one large.

Directions

1. Preheat oven to 450°F. Place two No. 6 Skillets on a baking sheet and transfer to preheated oven for at least 10 minutes.

2. In a small bowl, stir together strawberries, lemon juice, and 1 tablespoon sugar. Set aside.

3. Combine milk, ricotta, eggs, lemon zest, vanilla, sea salt, flour, and remaining 1 tablespoon sugar in the pitcher of a blender. Blend until smooth, 15–20 seconds. Keep batter at room temperature.

4. Remove skillets from the oven and add 1 tablespoon butter to each. Return to the oven until butter is melted and coats the inside, 2 minutes.

5. Remove skillets from the oven and pour half the batter into each skillet. Immediately return to the oven and bake until puffy and golden brown, 20 minutes. Important: Keep the oven door closed during baking! Opening the door will prevent the pancake from rising.

6. Remove pancakes from the oven. Spoon macerated strawberries over the top and dust with powdered sugar. Serve immediately.

Equipment

2 No. 6 Skillets by Smithey Ironware

blender

Ingredients

1 cup sliced strawberries

1 Tbsp. lemon juice

2 Tbsp. sugar, divided

⅓ cup whole milk

⅓ cup ricotta cheese

3 eggs

¼ tsp. lemon zest

½ tsp. vanilla extract

¼ tsp. fine sea salt

⅔ cup all-purpose flour

2 Tbsp. unsalted butter, divided

powdered sugar, for dusting

Courtesy Smithey Ironware.

Courtesy Stargazer.

Hearty, Spicy Shakshuka

Stargazer

The folks at Stargazer say this recipe is a flavor bomb! With Fresno peppers, Italian long hot peppers, and tomatillos, it's hot and spicy. If you want to cool it down a bit, substitute the vegetables and add a little sugar to cut the heat and acidity.

Directions

1. Preheat your oven to 375°F.

2. Preheat your Stargazer Cast-Iron Skillet on low for 5–10 minutes on the stove top before gradually turning the heat up to medium. Add the olive oil, onion, Fresno peppers, and long hot peppers. Sauté for 10 minutes, stirring regularly until softened.

3. Add the crushed tomatoes, tomatillos, half of the cilantro, garlic, cumin, paprika, salt, and pepper. Mix until the ingredients are thoroughly combined. Crack the eggs on top of the mixture one at a time, with space between them. To keep the eggs from running, make a little divot in the mixture with your spatula to crack the eggs into.

4. Transfer the skillet to the preheated oven. Bake for 20 minutes, just long enough to soften the veggies and cook the eggs while leaving them runny.

5. Remove the skillet from the oven and serve immediately over toast, sprinkling some of the remaining cilantro on top of each serving.

Equipment

Stargazer Cast-Iron Skillet

Ingredients

3 Tbsp. olive oil

1 medium onion, sliced thin

3 Fresno peppers, sliced in thin rounds

2 Italian long hot peppers (long hots), sliced in thin rounds

1 28-oz. can crushed tomatoes

3 tomatillos, sliced into 8 wedges each

½ cup chopped cilantro, leaves and stems separated, divided

3 cloves garlic, thinly sliced

¼ tsp. ground cumin

1 tsp. paprika

1 tsp. salt

½ tsp. ground black pepper

6 large eggs

6 slices of your favorite bread, toasted for serving

Cinnamon Apple Compote

Marquette Castings

The Marquette Castings family loves this simple apple compote on anything and everything. Who wouldn't?

Directions

1. In a bowl, mix together allspice, nutmeg, cloves, and cinnamon. Add the apple cubes and toss until covered in the spice mixture.

2. In your Marquette Castings Dutch Oven, melt the butter and brown sugar over low heat. Add cinnamon sticks and then the spiced apples. Simmer on medium, stirring occasionally, for 15–20 minutes. Remove cinnamon sticks and serve the apple compote over pancakes or oatmeal.

Equipment

Marquette Castings
Dutch Oven

Ingredients

1 tsp. allspice

1 tsp. nutmeg

1 tsp. cloves

2 Tbsp. cinnamon

5 apples, peeled and cubed

4 Tbsp. butter

4 Tbsp. brown sugar

2 cinnamon sticks

Breads

Crusty No-Knead Bread
Crisbee

The folks at Crisbee adapted this recipe from FrugalFitMom.com.
They say it's the easiest bread to make!

Directions

1. In a large bowl, stir all ingredients together and cover with
   plastic wrap. Let dough sit for 8–24 hours.

2. Remove dough and use floured hands and surface to form
   dough into a ball. Let rest for 30 minutes.

3. Preheat oven to 450°F with the Dutch oven inside. You can
   add a little bit of cornmeal or flour to the pot before adding
   the dough but do not oil the pot.

4. After dough has rested, remove pot from oven and add the
   dough. Bake covered for 30 minutes. Remove the lid and bake
   another 10 minutes. Slice and serve.

Equipment
cast-iron Dutch oven

Ingredients
3 cups all-purpose flour,
plus more for dusting

1 tsp. salt

½ tsp. yeast

1½ cups warm water

Cast-Iron Buttermilk Biscuits

Field Company

This classic recipe for buttermilk biscuits calls for shortening and butter to give it just the perfect balance of flakiness and softness.

Directions

1. Preheat the oven to 400°F.

2. In a medium bowl, whisk together the flour, baking powder, baking soda, and salt. Add the cold butter and shortening to the dry ingredients. Cut the butter and shortening into the flour until the mixture resembles coarse meal. Add buttermilk and stir gently with a large spoon until mixture just comes together. The dough will be sticky.

3. Transfer the dough to a floured cutting board and pat into a rectangle. Cut the dough into four equal pieces. Stack all four pieces on top of one another. Gently press down on the stacked pieces of dough, then roll into a rectangle that's about 1 inch thick.

4. Using a sharp knife, cut the dough into 8 equally sized squares.

5. Arrange the biscuits in a greased No. 8 Field Skillet so that they're just touching each other. Brush with butter and bake for approximately 15 minutes or until lightly browned.

6. Remove the biscuits immediately after baking and place on a clean kitchen towel to prevent the bottoms from cooking any further in the hot pan. Transfer the cloth and biscuits back to the skillet to keep warm until ready to serve.

Equipment

No. 8 Field Company Cast-Iron Skillet (10.25")

Ingredients

2 cups all-purpose flour, plus more for dusting

1 Tbsp. baking powder

½ tsp. baking soda

1 tsp. kosher salt

4 Tbsp. (½ stick) very cold butter, cut into ½-inch pieces

4 Tbsp. very cold vegetable shortening, cut into ½-inch pieces

1 cup cold buttermilk (low-fat or whole)

oil to grease pan

3 Tbsp. melted butter, for brushing

The Key to Better Biscuits

The folks at Field Company shared the following tips on making light, fluffy, perfectly flaky biscuits.

1. They have to be made in a cast-iron skillet. The pan's ability to retain heat gives the biscuits a lightly crisped bottom with the right amount of tender crumb.

2. Use both butter and shortening, and make sure they're completely chilled or frozen. The butter gives biscuits a flaky texture while the shortening makes the biscuits soft. If the butter is room temperature, though, it will melt and seep into the biscuit dough instead of allowing the water in the butter to evaporate as soon as the biscuits go into the oven.

3. Stack sections of your dough to give your biscuits those beautiful flaky layers.

4. Bake your biscuits at a fairly high temperature (like 400°F) to encourage evaporation inside the biscuits. This makes them rise quickly instead of spreading out.

5. Cast iron retains a lot of heat, so if you leave your biscuits in the pan after they're done, the bottom will continue to cook and may even burn. Instead, remove your biscuits and wrap them in a clean kitchen towel. Then place the wrapped biscuits back in the skillet to keep warm until you're ready to serve them.

Henry Lodge's Favorite Cornbread

Lodge Cast Iron

This cornbread recipe has been in the Lodge family for decades and is often made for family meals by former CEO Henry Lodge's wife. While there's no sugar in it, the addition of creamed corn brings a hint of sweetness to the savory bread. Plus, the corn and sour cream keep the cornbread moist and prevent it from drying out, making it just as good left over as it is fresh.

Directions

1. Preheat oven to 400°F.

2. While oven is preheating, pour ¼ cup of oil into a 10.25" Lodge Cast-Iron Skillet and place in oven to preheat.

3. In a bowl, mix remaining ingredients together. Remove skillet from the oven and pour out the hot oil into the cornmeal mixture. Stir together.

4. Pour cornmeal mixture into the hot skillet and bake for 35 minutes or until a tester comes out clean.

Equipment

10.25" Lodge Cast-Iron Skillet

Ingredients

½ cup canola oil, divided

1 cup self-rising cornmeal

1 8-ounce can creamed corn

3 eggs

1 cup sour cream

¼ tsp. salt

Ms. Judy giving a tour of a typical wood stove found in nineteenth-century homes in Florida. *Author's photo.*

Mexican Cornbread

Judith Stricklin

Judith Stricklin, a.k.a. Ms. Judy, is the living history interpreter at the Tallahassee Museum. On Saturdays, you'll find her in the old farmhouse's nineteenth-century kitchen slaving over a wood stove. While dressed in period garb, Ms. Judy is quick to provide a bit of history along with a bite of cookie or cornbread—all made with period-appropriate ingredients, of course. In the colder months, Ms. Judy often makes Mexican Cornbread to go along with a big pot of chili, which she serves to the workers and volunteers at the museum. Made with bacon, cheese, and cream corn, she claims one slice of this cornbread makes a meal. As an official taste-tester, I have to agree with her.

Directions

1. Preheat oven to 425°F.

2. In a bowl, mix dry ingredients until well combined. Add remaining ingredients and mix until moistened. Pour into a greased 12" skillet.

3. Bake 25–30 minutes or until done.

Equipment

12" cast-iron skillet

Ingredients

1½ cups yellow cornmeal

1 cup all-purpose flour

1 Tbsp. baking powder

1 tsp. salt

1 tsp. black pepper

1 cup buttermilk

½ cup oil, plus more to grease the skillet

3 eggs, well beaten

1 can creamed corn

1 cup cheddar cheese or Mexican blend, grated

3 (or to taste) minced hot peppers

½ onion, diced

3 slices crisp bacon, chopped

Turnip Green & Bacon Dip (with Pull-Apart Bread Crust)

American Skillet Company

Why does spinach have all the fun? This turnip green–based dip is a surprising crowd-pleaser, especially when served in an Oklahoma-shaped pan by American Skillet Company. The Oklahoma skillet holds three cups in volume, so you can substitute a 9" skillet if you need to.

Equipment

Oklahoma Skillet by American Skillet Company

Ingredients

4 slices bacon

¼ cup finely chopped yellow onion

5 cups coarsely chopped fresh turnip greens

½ package (8 oz.) cream cheese, softened

½ cup sour cream

½ cup shredded sharp white cheddar cheese, divided

¼ cup freshly grated Parmesan cheese

6 Tbsp. mayonnaise

¼ tsp. kosher salt

¼ tsp. garlic powder

¼ tsp. red pepper

⅛ tsp. ground cumin

1 can (11 oz.) breadsticks (Pillsbury Original Breadsticks works great)

Directions

1. Preheat oven to 375°F.

2. In your Oklahoma Skillet, cook bacon until crisp and then transfer it to paper towels to drain, reserving 2 tablespoons of the drippings in the skillet. Set aside bacon for later.

3. Add onion to hot skillet and cook until softened, 2 minutes. Add greens to the skillet and cook until wilted, 4 minutes. Transfer the greens and onions to a medium-size bowl and set the skillet aside.

4. To the greens and onion mixture, add cream cheese, sour cream, ¼ cup cheddar, Parmesan, mayonnaise, and the remaining seasonings.

5. Unroll dough onto a lightly floured surface, separating into 12 pieces. Roll each into a spiral bun, pinching seams to seal. Place rolls around the edge of the skillet, seam side facing out.

Photo by Marla Bergh, courtesy American Skillet Company.

6. Spoon the dip into the center of skillet. Bake until bread is golden brown and dip is bubbly, about 24 minutes, covering with foil during the last 4 minutes of baking to prevent top of dip from drying out.

7. Crumble bacon and sprinkle over bread and dip along with remaining ¼ cup of cheddar. Bake until cheese is melted, about 2 more minutes. Serve warm in the skillet.

Garlic-Parmesan Roasted Potatoes

Chef Lance

This simple potato recipe packs plenty of flavor, making it a great complement for hearty steaks.

Directions

1. Preheat oven to 400°F.

2. Lightly grease a large cast-iron skillet with bacon grease.

3. In a large bowl, add all potatoes, cheese, bacon, and seasoning, and toss thoroughly. Transfer potatoes to the skillet and bake for 45 minutes to 1 hour, turning halfway through.

4. Remove from oven and toss with melted butter before serving.

Equipment

large cast-iron skillet

Ingredients

2 lbs. Golden potatoes

¼ cup olive oil

1 Tbsp. garlic, thinly sliced

1 tsp. salt

½ tsp. rosemary

¼ cup Parmesan or Italian blend cheese

2–3 strips of cooked bacon, crumbled

red pepper flakes, to taste (optional)

1 Tbsp. butter, melted

Bacon-Wrapped Asparagus

Chef Lance

Asparagus is good by itself, but it's phenomenal wrapped in bacon. This easy side is perfect for date night.

Directions

1. Preheat oven to 400°F.

2. Partially cook bacon, making sure it's still flexible. Set on a rack or paper towels to drain.

3. Lightly oil a cast-iron skillet.

4. Trim asparagus bottoms. Depending on thickness, wrap 1–4 spears with one slice of bacon in a "barbershop" fashion. Place in the skillet in a single layer.

5. Bake the asparagus until done, 15 minutes.

Equipment

large cast-iron skillet or griddle

Ingredients

4 slices thick-cut bacon

oil to grease pan

½ lb. asparagus spears

Roasted & Seared Sweet Potatoes with Maple-Miso Glaze and Fried Sage

Field Company

This recipe allows sweet potatoes to get the most out of the oven but stops short of turning them to mush. Instead, the extra caramelization and flavor come from searing the potatoes, which are then glazed with a delicious sauce. As for frying the sage, this flavors the oil and gives you a crispy garnish for serving.

Directions

1. Preheat the oven to 425°F.

2. Arrange the whole sweet potatoes in a No. 12 Field Skillet. Roast the potatoes until just tender when pierced with a knife, 35–45 minutes. Transfer to a cutting board and let cool for at least 15 minutes.

3. While the potatoes are roasting, melt the butter in a small saucepan over medium-low heat. Add the maple syrup and miso and whisk until smooth. Turn off the heat and set aside.

4. Once the potatoes are done, add enough vegetable oil to the empty No. 12 Field Skillet to completely coat the bottom of the pan. (If you want to save oil, you can use a smaller skillet.) Heat the oil over medium-high heat until it shimmers. Add the sage leaves and fry, stirring constantly with a wooden spoon, until crisp, 10–15 seconds. Transfer the sage to paper towels and season with kosher salt.

5. Carefully remove all but about 3 tablespoons of the hot sage-infused oil from the No. 12 skillet. Slice the sweet potatoes

Equipment

No. 12 Field Skillet (13⅜" diameter)

saucepan

Ingredients

6 medium sweet potatoes (3 to 3½ lbs. total), scrubbed

4 Tbsp. (½ stick) unsalted butter

¼ cup maple syrup

¼ cup white miso

vegetable or grapeseed oil, for frying the sage

20 large sage leaves

kosher salt

flaky sea salt

Courtesy Field Company.

crosswise, skin and all, into 2-inch pieces. Set the skillet over medium-high heat. When the oil is shimmering, add the sweet potatoes, cut side down, and cook until deeply browned on the bottom, 2–3 minutes. Turn the sweet potatoes over and cook until the other side is well browned. Pour out excess oil.

6. Pour the butter-maple mixture over the sweet potatoes and cook, basting the potatoes with a spoon until the glaze has thickened and coated the potatoes, 1–2 minutes. Transfer the sweet potatoes to a serving platter and drizzle with any remaining glaze. Sprinkle the sweet potatoes with flaky salt and scatter the fried sage over the top. Serve.

Photo by Eva Kosmas Flores, courtesy FINEX.

5-Ingredient Bacon Caramelized Brussels Sprouts

FINEX

FINEX contributed this simple recipe from Eva Kosmas Flores of Adventures in Cooking. They say this rendition of the tried-and-true bacon and Brussels sprouts is hard to beat and worth bookmarking for future cast iron adventures.

Directions

1. Place a FINEX 10" skillet over medium heat. Add the chopped bacon and sauté until slightly cooked but not crispy. Remove from pan and set aside.

2. Add the remaining ingredients to the pan and sauté until the Brussels sprouts soften and turn a brighter and deeper shade of green, about 10 minutes, stirring every 2–3 minutes.

3. Add the bacon back to the pan and stir to combine. Continue cooking until the bacon is crispy and the Brussels sprouts have turned gold around the tips and are relatively soft when pierced with a fork, about 6–8 minutes more, stirring every 2 minutes. Pairs well with a hearty meat like pork chops.

Equipment

FINEX 10" Skillet

Ingredients

5 thick bacon slices, cut into ½" squares

1 lb. Brussels sprouts, each sprout cut in half

2 Tbsp. olive oil

1 tsp. salt

½ tsp. black pepper

Ratatouille

Stargazer

Ratatouille is one of those dishes that's just as pretty as it is tasty. Serve it in the skillet for the ultimate wow factor.

Directions

1. Preheat oven to 375°F.

2. In your Stargazer Skillet, heat a bit of olive oil then add minced garlic, red pepper, and onion and cook until soft. Pour in crushed tomatoes and cook for 10 minutes.

3. Add remaining vegetables to the skillet, arranging them in a circular pattern. Drizzle with olive oil and sprinkle with salt and pepper.

4. Bake covered (with lid or foil) for 20 minutes. Remove cover and bake for another 10–15 minutes. Sprinkle with chopped parsley and serve.

Equipment

10" or 12" Stargazer Skillet

Ingredients

2 Tbsp. olive oil, divided

2 gloves garlic, minced

1 red pepper, cubed

½ white onion, chopped

28 oz. can crushed tomatoes

2 zucchinis, sliced into ⅛" thick rounds

2 yellow squash, sliced into ⅛" thick rounds

5 Roma tomatoes, sliced into ⅛" thick rounds

2 long eggplants, sliced into ⅛" thick rounds

1 tsp. salt

cracked pepper

fresh parsley, chopped

Courtesy Stargazer.

Twice-Baked Potato Casserole

Lodge Cast Iron

Transform the classic twice-baked potato into a cheesy bacon-topped casserole in this comfort-food recipe from Lodge Cast Iron.

Directions

1. Preheat oven to 400°F.

2. With a fork, poke holes in potatoes and then wrap individually in foil. Bake until tender, about 1 hour.

3. After potatoes are done, remove from foil, cut in half, and scrape the potato out of the skin into a large bowl. Add sour cream, heavy cream, milk, cream cheese, melted butter, egg, garlic, salt, and pepper to potatoes and stir to combine. Then add ¼ cup scallions and ½ pound of cooked bacon and mix well. Spread potato mixture evenly into a 3.6 Quart Casserole Dish. Top with cheddar cheese and remaining bacon.

4. Bake potato casserole, rotating once, until top is slightly browned, 20–30 minutes. Garnish with remaining scallions and serve.

Equipment

3.6 Quart Enameled Casserole Dish by Lodge Cast Iron

Ingredients

12 medium potatoes

¾ cup sour cream

½ cup heavy cream

1½ cups milk

¾ cup cream cheese

6 Tbsp. unsalted butter, melted

1 egg, beaten

1 Tbsp. minced garlic

½ tsp. salt

½ tsp. pepper

½ cup chopped scallions, divided

1 lb. cooked bacon, chopped and divided

½ cup shredded cheddar cheese

Main Dishes

Detroit Style Pizza

American Skillet Company

This Detroit, Michigan–inspired pizza was adapted by Jonny Hunter of the Underground Food Collective from Kenji Alt-Lopez's "Serious eats foolproof pizza" recipe. This recipe calls for two Michigan Skillets from American Skillet Company, but you can use two 9" skillets or one large skillet if you need to.

Directions

1. Combine flour, salt, yeast, water, and oil in a large bowl, mixing until no dry flour remains.

2. Cover bowl tightly with plastic wrap, making sure that edges are well-sealed, then let rest on the counter for at least 8 hours and up to 24. Dough should rise dramatically and fill bowl.

3. Flour work surface and transfer dough to it. Form dough into two balls.

4. Pour 2 tablespoons of oil in the bottom of each Michigan Skillet. Place 1 ball of dough in each pan and turn to coat evenly with oil. Press the dough around the pan until the entire bottom and edges of the pan are covered. Cover each pan with plastic wrap and let the dough sit at room temperature for 2 hours. After 2 hours, dough should mostly fill the pan up to the edges. Press it until it fills every corner of the pan.

Equipment

2 Michigan Skillets from American Skillet Company

Ingredients

1 cup bread flour

1 tsp. kosher salt, plus more for sprinkling

½ tsp. instant yeast

¾ cup water

1½ tsp. olive oil, plus more to coat pans and drizzle

1 cup pizza sauce, divided

6 oz. full-fat dry mozzarella cheese, grated

pepperoni (optional)

1 oz. Parmesan or Pecorino Romano cheese, grated

Photo by Marla Bergh, courtesy American Skillet Company.

5. Preheat oven to 550°F. Cover the dough with ¾ cup sauce. Top with mozzarella cheese, making sure the cheese goes all the way to the edges. Season with salt. Add pepperoni and other toppings as desired. Drizzle with olive oil.

6. Transfer the skillet back to the oven and bake for 12–15 minutes. The bottom should be crisp and the cheese should be golden. Sprinkle with grated Parmesan or Pecorino Romano cheese. Serve hot, right out of the skillet.

Award-Winning Chili

Cast Iron Steve and his brother Migration Mike

Cast Iron Steve and his brother, Migration Mike, have won three awards for this spicy and hearty three-meat chili. Their trick is to use a pellet smoker to give the peppers and onions a smoky flavor, but if you need to, you can roast yours in the oven at 350°F until soft.

Directions

1. Preheat pellet smoker to 350°F. Roast peppers and onions on pellet smoker until peppers start to blister and onions start to get soft.

2. Remove skin from peppers and cut up peppers and onions. Set aside.

3. Put bacon on pellet smoker and cook until done. Cut into small pieces.

4. In a large Dutch oven, brown sausage. Add smoked peppers and onions and cooked bacon as well as tomatoes, tomato sauce, beef stock, and brisket. Let simmer for 1–2 hours.

5. Add drained beans and corn to Dutch oven. Then add salt and chili powder to taste (it won't need much). Let simmer for another 1–2 hours.

6. Serve as is or garnish with sour cream, cheese, crispy onions, and green onions.

Equipment

pellet smoker (or pans to roast in oven)

large Dutch oven

Ingredients

1 jalapeno pepper

2 poblano peppers

2 Anaheim peppers

1 each red, yellow, and orange bell peppers

2 yellow onions

½ lb. bacon

1 lb. sausage

2 cans fire-roasted tomatoes

1 can tomato sauce

32 oz. beef stock

2–3 lbs. cooked brisket (tri-tip or prime rib could also be used)

1 can pinto beans, drained

1 can black beans, drained

1 can kidney beans, drained

2 cups corn, fresh or frozen

salt to taste

chili powder to taste

sour cream, shredded cheese, crispy onion, and green onions for serving

Courtesy Cast Iron Steve, pictured on right.

Chesapeake Shore Bird

Butter Pat Industries

The folks at Butter Pat Industries say nothing gives chicken a crispy skin with juicy meat like the combination of cast iron and this magic brining/drying recipe. The original method comes from the British chef Fergus Henderson but is adapted here to use buttermilk and a little Baltimore fish pepper sauce in the brine.

You're welcome to substitute the chicken with any poultry or wild game bird. Just remember to vary your brine and cooking times to compensate for the weight of the bird. You can also roast root vegetables, mushrooms, or apples along with the meat. This recipe requires two days advance prep, though, so plan ahead—it's worth it!

Directions

1. To make the brine, add water, salt, bay leaves, thyme, oregano, and pepper flakes to a pot and boil to dissolve the salt. Allow brine to cool to room temperature, then add buttermilk and fish pepper sauce. Put brine and chicken in a large plastic bag and place in the refrigerator for 4–8 hours. The longer the poultry is in the brine, the saltier it will be.

2. Remove the chicken from the brine and dry completely with paper towels—do not rinse. Place the chicken back in the refrigerator on a wire rack in a sheet pan, uncovered, for 12–24 hours. This step will dry the skin, making it crispy, but the meat will remain juicy.

Equipment

pot to boil brine

wire rack

baking pan

14" Lili or 12" Joan Cast-Iron Skillet by Butter Pat Industries

Ingredients

whole chicken, backbone removed, and breast broken (spatchcock)

pepper, to taste

onion, halved

For brine

4 cups water

½ cup kosher salt

2 bay leaves

½ tsp. thyme

½ tsp. oregano

¼ tsp. pepper flakes, optional

4 cups buttermilk

1 Tbsp. Snake Oil fish pepper sauce or equivalent hot sauce, optional

For herb butter

3 garlic cloves, minced

¼ tsp. thyme

¼ tsp. oregano

3 Tbsp. butter, melted

3. Prepare the herb butter in a separate bowl by adding garlic, thyme, and oregano to melted butter. Place in refrigerator to solidify.

4. To cook the chicken, start by seasoning with pepper on all sides. With your fingers, scoop small gobs of the herb butter and place between the skin and the meat in as many places as possible. Melt and retain the leftover herb butter for the cooking baste. Leave the bird at room temperature while preheating the oven to 500°F. Place the 14" Lili or 12" Joan cast-iron pan into the oven to preheat.

5. Once the oven has reached 500°F, remove the pan and add onion halves. Open the bird and place upside down—chest cavity side up—over the onions. Roast the bird for 10 minutes.

6. Turn the bird breast side up. Baste the breast with reserved butter and return it to the oven for 10 minutes.

7. Remove the pan and lower the oven temperature to 350°F. Allow the oven to cool until it reaches temperature. Then return the pan to the oven and roast the chicken until the juices run clear and internal temperature is 155–165°F, 20–30 minutes.

8. Remove the pan from the oven and tent it with aluminum foil. Allow to rest 10–15 minutes. This is a critical step as the bird will continue to cook during the rest. If you skip this step, the juices will run out when you cut the bird. Wait for perfection, then serve.

Braised Chicken Thighs with Garlic, Olives, and Lemon

Field Company

Keep this highly adaptable recipe in your back pocket for any night you want a quick one-pot meal. This dish has a Mediterranean flair with garlic, olives, and rosemary, but you can easily globetrot to Asia (ginger, sesame, soy), Mexico (chiles, peppers, lime), France (white wine, tarragon, Dijon), Italy (tomatoes, basil, black olives), or anywhere your pantry guides you.

Directions

1. Heat the oil in a No. 8 (10¼") cast-iron skillet over medium-high heat until it begins to shimmer. Pat the chicken skin dry with paper towels, then season with salt and pepper. Add the chicken to the skillet and cook, without disturbing, until the skin is golden brown, 5–7 minutes. Turn the chicken over and add the garlic. Cook until the chicken is browned on the other side and the garlic is golden brown, 5 minutes. Transfer the chicken and garlic to a plate.

2. Add the onion to the skillet, lower the heat to medium, and cook, stirring, until the onion is softened and beginning to brown, 8–10 minutes. Return the chicken and garlic to the skillet and add the olives, rosemary, bay leaf, and lemon zest. Add enough chicken stock to cover most of the chicken, leaving the skin exposed. Bring the liquid to a simmer, then reduce the heat to low, cover the skillet, and braise until the chicken is cooked through, 30 minutes.

3. Stir the lemon juice into the sauce and season to taste with salt and pepper. Sprinkle with parsley and serve.

Equipment

No. 8 Field Skillet (10¼") with fitted lid

Ingredients

1 Tbsp. vegetable oil

6 bone-in chicken thighs with skin

kosher salt, to taste

freshly ground black pepper, to taste

4 garlic cloves, smashed

1 medium onion, thinly sliced

½ cup pitted green olives, halved

1 sprig rosemary

1 bay leaf

4 strips of lemon zest

1–2 cups chicken stock

1 Tbsp. fresh lemon juice

¼ cup parsley, chopped

Courtesy Austin Foundry Cookware.

Southwestern Flat Iron Steak Skillet Delish

Austin Foundry Cookware

Flat iron steaks—also known as top blade steaks or top blade filets—have a nice marbling with plenty of beef flavor. They're a good alternative to more expensive steaks, cooking up tender and juicy. This is an easy recipe with plenty of Southwestern flavor, making it perfect for any night of the week.

Directions

1. Marinade steak for an hour with a mixture of spices and lime juice.

2. Preheat your AFC cast-iron skillet over medium-high heat. Add 1 tablespoon olive oil to skillet and turn heat to high. Place marinated steak into skillet, cooking each side until nicely charred or internal temp reaches 120–125°F, 2–3 minutes. Remove steak and slice into thin strips against the grain.

3. Put sliced steak back into skillet over medium-high heat. Add remaining ingredients and top with cilantro. Cook 2–3 minutes, constantly tossing. Serve.

Equipment

large cast-iron skillet from Austin Foundry Cookware

Ingredients

flat iron steak (or steak of your choice)

spices: salt, pepper, cumin, smoked paprika, cilantro

lime juice

olive oil

lime wedges

cherry tomatoes

salsa verde sauce

PRO TIP Preheat your cast iron for 5–10 minutes on low before adding oil or food or increasing the heat. This helps you avoid rapid temperature changes that can warp metal. According to Stargazer, "Preheating your cast iron cookware will also ensure that your food hits a well-heated cooking surface, which prevents it from sticking and aids in nonstick cooking."

Perfect Ribeye with Mushrooms and Onions

Chef Lance

This easy steak has tons of flavor, thanks to the "dry age" method, plenty of butter and seasoning, and a mouthwatering topping of caramelized onions and mushrooms.
You just can't go wrong with this steak!

Directions

1. Liberally season steak on both sides, rubbing the seasoning in thoroughly. Place on a wire rack on a baking sheet and place in refrigerator for 24–48 hours for a quick "dry age." Do not cover the steaks.

2. Remove steaks from refrigerator 45 minutes to an hour before cooking and allow to come to room temperature.

3. Heat a large skillet on medium-high heat. Add a generous amount of butter to the hot pan. When it's melted and sizzling, add the steak. Cook for about 2–3 minutes on the first side. Reduce heat to medium-low, flip steak, and finish cooking for another 2–3 minutes, basting with pan fat. This will result in medium-rare steak. For more flavor, add balsamic vinegar, red wine, or soy sauce during cooking.

Equipment

wire rack

baking sheet

large cast-iron skillet

Ingredients

12 oz. rib eye steak (at least 1" thick)

2 tsp. seasoning of choice, including salt and pepper

1 Tbsp. oil

several mushrooms, sliced

½ onion, sliced

1–2 Tbsp. butter

balsamic vinegar, red wine, or soy sauce (optional)

4. Optional—If you prefer your steak cooked more thoroughly, place the seared steak in the oven at 300°F for just a few minutes. Check the temperature every couple minutes to ensure it doesn't overcook.

5. Remove cooked steak from the pan and set on cutting board. Allow the steak to rest for 10 minutes before cutting.

6. While steak rests, place pan back on burner on low to medium heat. Add mushrooms and onions. Sauté until done, 7–10 minutes.

7. Serve the steak topped with sautéed mushrooms and onions.

Grilled & Reverse-Seared Steak

Field Company

A popular way to cook thick steaks is to sear them on the stove and then finish them in the oven. With a reverse-seared steak, though, you start with gentle heat until the steak reaches a target temperature then blast it with high heat for a burnished crust. This recipe puts a twist on the reverse-sear method by basting the meat in a mixture of melted butter, garlic, and herbs while it's searing.

Directions

1. Season the steak generously with salt and place on a wire rack set inside a rimmed baking sheet. Refrigerate, uncovered, for 4–12 hours.

2. Prepare a two-stage grill with hot and cool sides. Set the steak over the cool side and cook uncovered, turning occasionally, until the center registers 105°F for rare, 115°F for medium-rare, or 125°F for medium. Depending on your grill, this can take anywhere from 15–30 minutes. Don't cover the grill, as that would cook the steak too quickly. Transfer the steak to a plate.

3. Set a No. 8 Field Skillet over the hot side of the grill and let it heat up for a few minutes. Add the butter, garlic, and herbs. When the butter foams, add the steak and cook, basting constantly with the butter, until well browned on the bottom, 30–45 seconds. Turn the steak over and cook, basting with butter, until the other side is browned. Transfer the steak to a plate and pour the contents of the skillet over it. There's no need to let the steak rest with this method, so carve and serve right away.

Equipment

wire rack

baking sheet

outdoor grill

No. 8 Field Skillet (10.25") or larger

Ingredients

1 rib eye, 1½–2 inches thick

kosher salt

4 Tbsp. (½ stick) unsalted butter

3 garlic cloves, peeled and smashed

1 sprig rosemary and/or a small bunch of thyme

Cast-Iron-Seared Standing Rib Roast

Lodge Cast Iron

Impress your guests with a fall-off-the-bone roast and flavorful veggies with this standing rib roast recipe.

Directions

1. Place the rib roast on a wire rack inside a rimmed baking sheet and coat with salt. Refrigerate overnight, up to a day.

2. Preheat oven to 450°F.

3. Puree garlic and shallots, then mix with Dijon, 1 tablespoon olive oil, rosemary, oregano, thyme, and cracked pepper. Set aside.

4. Preheat empty skillet over medium-high heat for 5 minutes. Brush off excess salt from the roast and sprinkle with pepper. Add remaining 2 tablespoons of olive oil to the skillet, and sear the roast on all sides, 3–4 minutes per side. Remove the roast from the skillet.

5. Add carrots, celery, and onion and toss to coat in beef drippings. Return the roast to skillet (with roast on top of veggies) and brush with garlic and shallot mixture. Place the skillet in the oven and roast for 25 minutes.

6. Lower the oven temperature to 350°F and continue roasting until the internal temperature reaches 135°F, 1 hour and 35 minutes (about 16 minutes per pound). Remove from the oven and let rest for 15 minutes before serving. Slice and serve atop veggies, spoon pan drippings over top, and, if you like, serve with horseradish.

Equipment

wire rack

baking sheet

12" Chef Collection Skillet by Lodge

Ingredients

1 bone-in standing rib roast (5–6 lbs.)

1 cup kosher salt

5 cloves garlic, chopped

2 shallots, chopped

1 Tbsp. Dijon mustard

3 Tbsp. olive oil, divided

2 Tbsp. rosemary

1 Tbsp. oregano

2 tsp. thyme

2 tsp. cracked pepper

4 large carrots, peeled and cut into 1-inch pieces

4 large stalks celery, cut into 1-inch pieces

2 medium red onions, chopped

horseradish sauce for serving

Courtesy Lodge Cast Iron.

Braised Short Ribs & Cabbage

Smithey Ironware

This recipe is made for Dutch ovens. Cooked low and slow, the braised short ribs come out tender and the cabbage flavorful. This is a perfect dish for a cool night.

Directions

1. Preheat oven to 300°F.

2. Season short ribs generously on all sides with salt and pepper. Heat canola oil in 3.5-quart Dutch oven over medium-high heat. Add half the short ribs to the oil and sear until deeply brown and caramelized on all sides, 10 minutes. Repeat with remaining short ribs. Transfer ribs to a plate.

3. Carefully discard all but 1 tablespoon fat from the pot and reduce heat to medium. Add onions, leeks, and garlic and cook until softened, 5–8 minutes. Stir in tomato paste and paprika and cook for 2 minutes. Stir in wine and cook 2 minutes longer.

4. Add beef broth and bring to a simmer. Add cabbage in batches, letting each addition wilt gently to make room for more. Return seared short ribs to the pot and season with salt and pepper to taste.

5. Cover and transfer Dutch oven to the preheated oven. Bake until short ribs are tender and just about to fall off the bone, 2½–3 hours. Remove lid during the last 30 minutes of cooking to allow sauce to reduce.

Equipment

3.5-Quart Dutch Oven by Smithey Ironware

Ingredients

3 lbs. bone-in short ribs

coarse kosher salt, to taste

freshly ground black pepper, to taste

1 Tbsp. canola oil

2 cups onion, thinly sliced

1 cup thinly sliced leeks, washed and spun dry

3 garlic cloves, thinly sliced

1 Tbsp. tomato paste

1 tsp. smoked paprika

¾ cup dry white wine

2 cups beef broth

1 lb. cabbage, cut into large dice (about half a medium-size head)

mashed potatoes or rice for serving

For gremolata

⅓ cup parsley, finely chopped

2 tsp. garlic, minced

¼ cup horseradish, freshly grated

Courtesy Smithey Ironware.

6. Meanwhile, make the horseradish gremolata. In a small bowl, stir together parsley, garlic, and grated horseradish.

7. Spoon stew into bowls on top of mashed potatoes or rice. Garnish with a few spoonfuls of horseradish gremolata.

Skillet-Stuffed Shells with Italian Sausage

Stargazer

Packed with garlic, Parmesan, and ricotta cheese, and then topped with Italian sausage, this flavorful pasta dinner is sure to please the whole family (while sneaking in a bit of spinach).

Directions

1. Preheat oven to 350°F.

2. Bring a pot of water to a boil and cook the shells to al dente according to the directions, then drain and set aside.

3. Remove Italian sausage from casing and cook, while crumbling, in 12" Stargazer Cast Iron Skillet. Remove from heat, drain the fat, and scoop sausage into a bowl. Carefully wipe out the skillet with a paper towel.

4. In a large bowl, combine 10 oz. of the shredded mozzarella, ricotta, ½ cup of the Parmesan, garlic, eggs, herbs, and spices. Then stir in the chopped spinach.

5. Coat the bottom of both skillets with spaghetti sauce. Spoon the ricotta mixture into each shell, placing them open side up and closely together in each skillet.

Equipment

pot to boil pasta

12" Stargazer Cast Iron Skillet

10.5" Stargazer Cast Iron Skillet

Ingredients

12 oz. package jumbo pasta shells

16 oz. Italian sausage (mild or hot)

16 oz. shredded mozzarella cheese, divided

32 oz. ricotta cheese

¾ cup grated Parmesan cheese, divided

5 cloves fresh garlic, chopped

2 eggs, lightly beaten

3 tsp. garlic powder

1 tsp. dried oregano

1 tsp. dried basil

½ tsp. salt

½ tsp. pepper

2 cups spinach, roughly chopped

2 24-oz. jars of marinara sauce

crushed red pepper and additional Parmesan cheese for serving

Courtesy Stargazer.

6. Add the Italian sausage crumbles over the top of the shells then evenly sprinkle with the leftover 6 oz. of shredded mozzarella and ¼ cup of Parmesan. Cover each skillet with aluminum foil and bake for 30 minutes. Remove the aluminum foil and switch the oven to broil, moving the skillets to the top rack and broiling until the cheese browns, 3–5 minutes. Remove from the oven and allow to cool for 5 minutes before serving.

Kent Rollins says not to boil water in seasoned cast iron because some of the seasoning will come loose and discolor the water. Use aluminum, stainless steel, or enamelware instead.

Courtesy Austin Foundry Cookware.

Blackened Chilean Sea Bass

Austin Foundry Cookware

Like many of us, Sean and Lisa Girdaukas of Austin Foundry Cookware cook primarily "by feel." They may write down a list of ingredients, but they don't keep track of the exact quantity as they cook. As you'll see with this simple but delicious recipe for sea bass, that's not a problem. Cook as many pieces of fish as your family requires and season to taste. Voila!

Directions

1. Pat dry the sea bass and dust it with your favorite Cajun seasoning.

2. Heat your AFC cast-iron skillet over medium-high heat on stovetop. Add butter. Place sea bass in the skillet and cook each side until flaky or internal temperature is 145°F, 4 minutes. Remove fish from skillet.

3. In the empty skillet, add a bit of olive oil, lemon slices, and cherry tomatoes. Stir while cooking until the tomatoes start to blister. Turn off heat. Add salad greens to skillet and toss with olive oil, salt, pepper, and a sprinkle of red wine vinegar. Add fish back to pan and serve.

Equipment

large cast-iron skillet from Austin Foundry Cookware

Ingredients

sea bass

Cajun seasoning

2 Tbsp. butter

olive oil

lemon slices

cherry tomatoes

salad greens

salt

pepper

red wine vinegar

Parmesan-Crusted Salmon in Creamy White Wine Dill Sauce

Stargazer

Every home cook loves a meal impressive enough to serve on special occasions while also simple enough for an easy midweek dinner. This recipe is exactly that with a perfect combination of flaky, buttery fish and crispy, cheesy crust, served over your favorite pasta.

Directions

1. Preheat oven to 375°F.

2. Preheat a large Stargazer skillet on low heat for 5–10 minutes.

3. Lightly coat the salmon fillets with oil, then season with salt and pepper. Once the skillet is preheated, add cooking oil and gradually increase the temperature to medium-high. Add the salmon fillets to the hot skillet and sear until lightly browned, 1 minute on each side. Remove the salmon and set aside.

4. Deglaze the pan by adding the white wine and scraping off the fatty residue from the pan using a spatula. This will allow the fat to season the liquid, creating a perfect base for your sauce.

5. Add the heavy cream to the skillet and bring to a boil. Reduce to a simmer and slowly add the flour while stirring continuously to prevent clumping. Add salt, pepper, dill, and capers. Continue to simmer and stir constantly until the sauce thickens to a thin gravy consistency, 5–7 minutes.

6. Add the seared salmon back to the skillet and drizzle some of the pan sauce over fillets with a spoon. Sprinkle shredded

Equipment

large skillet by Stargazer

pot to boil pasta

Ingredients

2 lbs. fresh salmon or about half a full salmon fillet, cut into 4–6 oz. sections

1 Tbsp. salt

½ Tbsp. pepper

3 Tbsp. high-heat cooking oil (ex. grapeseed, canola, avocado)

2 cups dry white wine

1 cup heavy cream

3 Tbsp. all-purpose flour

⅓ cup fresh dill, minced

4 Tbsp. pickled capers

½ cup Parmesan cheese, freshly grated

1 lb. fettuccine (or your favorite pasta)

dill sprigs for garnish

Courtesy Stargazer.

Parmesan cheese over the entire dish, making sure to coat the fillets.

7. Place the skillet in the preheated oven and cook for 15 minutes.

8. During this time, prepare your pasta al dente. This will prevent it from becoming overcooked and mushy once combined with the heated sauce.

9. Remove the skillet from the oven and let cool for 5 minutes. Serve over pasta, garnished with dill sprigs. Pair with a nice loaf of sourdough bread, any roasted vegetable, and your favorite white wine for maximum enjoyment!

PRO TIP Always preheat your skillet on low heat for 5–10 minutes. This prevents food from sticking and protects your skillet from warping. —Stargazer

Thanksgiving Turkey & Gravy

Fired Cast Iron—Dee Holbrook

Dee Holbrook cooks in a vintage style, often in a hearth or over a wood stove. This recipe has all the vintage appeal of a beautiful roasted turkey with the modern convenience of an oven roasting bag. Make sure you use a cast-iron brazier or Dutch oven that is large enough to hold the bird and the drippings it will produce.

Directions

1. Remove neck and giblets from the thawed turkey. Wash and dry the turkey thoroughly. Fill the body cavity with carrots, celery, a quartered onion, and garlic cloves (or any aromatic vegetable you wish). Then place the fresh herbs into the body cavity with the vegetables. Rub the outside of the turkey with butter or margarine, then generously salt and pepper.

2. Take the oven roasting bag and pour the flour inside. Hold bag closed and shake well to coat the inside of the bag with the flour. This prevents the bag from exploding during the baking process. Being careful not to rub off the butter and spices, place the turkey into the bag, breast side up. Add the bouillon cubes to the bag, all around the turkey. Close up the bag and cut three or four one-inch slits on top of the bag for ventilation. Place in a large enameled cast-iron brazier. Bake according to the directions provided with the oven roasting bag.

3. After the turkey has baked and has been removed from the oven, make the gravy. Pour the turkey drippings into a large pan over medium-high heat and bring to a boil. In a bowl,

Equipment

oven roasting bag

large enameled cast-iron brazier (or seasoned brazier or Dutch oven)

pan for gravy

Ingredients

15–22 lb. turkey, completely thawed

carrots, peeled and cut

celery sticks, cut

red onion, quartered

garlic cloves

rosemary, to taste

thyme, to taste

sage, to taste

butter or margarine

salt and black pepper

10 Wyler's chicken bouillon cubes or similar brand

2 Tbsp. all-purpose flour

For gravy

cornstarch

black pepper, to taste

milk

Courtesy Fired Cast Iron.

make a thickening mixture of cornstarch, black pepper, and milk, mixing well to remove clumps. Once the drippings are boiling, add the thickening mixture while stirring vigorously with a whisk. Continue stirring over medium-high heat until the gravy has thickened to a creamy consistency. Add pepper to taste. Serve turkey and vegetables with gravy on the side.

Sweets

Cinnamon Roll Coffee Cake

Crisbee

Traditional coffee cake meets cinnamon roll. The folks at Crisbee adapted this yummy breakfast treat from cakescottage.com.

Directions

1. Preheat oven to 350°F.

2. In a large bowl, mix cake ingredients together until well combined. Pour into a greased cast-iron Bundt pan or whatever cast-iron piece you prefer.

3. For the topping, mix ingredients together in a small bowl until well combined. Spread evenly on the batter and swirl with a knife. Be sure to swirl down into the batter for the "cinnamon roll effect." Bake for 30–35 minutes.

4. For the glaze, mix ingredients together in a small bowl until smooth. Drizzle the glaze over the warm cake. Serve cake warm or at room temperature.

Equipment

Cast-Iron Bundt Pan (or loaf pan or skillet)

Ingredients

For cake

½ cup butter, melted

2 eggs

¾ cup sugar

3 cups all-purpose flour

2 tsp. vanilla

4 tsp. baking powder

¼ tsp. salt

1½ cups milk

For topping

1 cup butter, melted

1 cup brown sugar

2 Tbsp. all-purpose flour

1 Tbsp. cinnamon

¼ chopped walnuts (optional)

For glaze

1 cup powdered sugar

2½ Tbsp. milk

½ tsp. vanilla

Fried Cinnamon Sugar Donuts

Austin Foundry Cookware

Directions

1. In a large bowl mixer, mix the sugar, baking powder, cinnamon, salt, egg, milk, and butter. Add 1½ cups flour and beat on low/medium for 2–3 minutes. Then add remaining flour and beat for additional 2 minutes. Wrap the dough and place it in the refrigerator for one hour.

2. While the dough chills, make the topping by combining the sugar and cinnamon in a separate bowl. Set it aside.

3. Pour 1½–2 inches of oil in your AFC cast-iron skillet and heat it to 360°F.

4. While skillet is heating up, roll dough to half-inch thickness on a floured surface and cut out with a donut cutter. Drop the donuts in the oil one or two at a time, flip them when they puff up, and pull them out when golden brown. Drain on a paper-lined dish.

5. Roll the warm donuts in the sugar-cinnamon mixture and serve immediately.

Equipment

bowl mixer

large cast-iron skillet from Austin Foundry Cookware

donut cutter

Ingredients

Donuts

½ cup sugar

2 tsp. baking powder

¼ tsp. cinnamon

pinch of salt

1 large egg

½ cup milk

2 Tbsp. melted butter

2 cups all-purpose flour, divided

canola or vegetable oil for frying

Topping

3 Tbsp. granulated sugar

1 Tbsp. cinnamon

Courtesy Austin Foundry Cookware.

Grandma's Apple Crumb Pie

Crisbee

Within the Crisbee family, a certain Grandma Cauble was known for making this delicious pie using a recipe she clipped from a *Better Homes and Gardens* magazine from October 1955. It's been a family favorite ever since.

Directions

1. Preheat oven to 400°F.

2. Place the pie crust into a greased skillet.

3. Peel and slice the apples then layer them onto the crust. Combine the sugar and cinnamon and sprinkle it over the apples.

4. In a bowl, make the topping by combining the remaining sugar and flour. Cut in the butter until crumbly. Sprinkle the topping over the apples.

5. Bake the pie for 40 minutes or until done.

Equipment

9" cast-iron skillet

Ingredients

Pie

9" unbaked pie crust

oil or shortening to grease skillet

5–7 golden delicious apples (gala and McIntosh work well, too)

½ cup sugar

1 tsp. cinnamon

Topping

½ cup sugar

¾ cup all-purpose flour

⅓ cup butter

Grilled Ginger Asian Pear Crisp
FINEX

The folks at FINEX shared this recipe from Lauren Chandler, proving that grill pans aren't just for burgers! You can use this recipe with any sort of seasonal fruit, including apples, strawberry rhubarb, or summer berries. This recipe can also be made using nondairy butter, gluten-free flour, and coconut or date sugar.

Directions

1. Preheat oven to 350°F. Oil the grill pan and place it in the oven while it is preheating.

2. In a large bowl, toss together the Asian pears, olive oil, ginger, and salt until well mixed.

3. Carefully remove the grill pan from the oven and place it on the stove over medium heat. Cook the fruit on both sides until you see grill marks. You will need to do this in batches, so keep the bowl next to the stove to transfer the cooked fruit. Re-oil the grill in between batches. Once all of the fruit is grilled, turn off the stove and transfer the fruit back to the grill pan in an even layer.

4. In a food processor, combine all of the topping ingredients and pulse until the mixture forms coarse crumbs. Spread the mixture evenly over the fruit and bake until the topping is lightly browned and crisp, 40 minutes. If you like a darker crisp topping, broil on the top rack for 3–4 minutes. Serve immediately with vanilla ice cream.

Equipment

FINEX 12" Grill Pan

food processor

Ingredients

Pears

oil for the pan

8–10 Asian pears, sliced into ½-inch half-moons

3 Tbsp. extra virgin olive oil, plus more for the grill pan

2 Tbsp. ginger, finely minced or grated

½ tsp. salt

Topping

½ cup all-purpose flour

1 stick cold unsalted butter, cut into 8 pieces

½ cup packed brown sugar

2 Tbsp. sugar

1 tsp. sea salt

2 cups chopped pecans and/or walnuts

1 cup rolled oats

vanilla ice cream for serving

Photo by Dana Halferty, courtesy FINEX.

Skillet S'mores

American Skillet Company

This s'mores dip is perfect for any gathering. The recipe calls for a USA Skillet by American Skillet Company, which holds 2 cups in volume. If you're still waiting for your USA Skillet to come in, you can use an 8" or 9" cast-iron skillet.

Directions

1. Turn the oven to medium broil with the rack in the middle position.

2. While the oven is heating, place your USA Skillet on a burner over low heat. Melt the butter and use it to coat the bottom and sides of your pan.

3. Once the butter is melted, toss in a "healthy layer" of chocolate chips to cover the bottom of the skillet, then place one layer of marshmallows on top.

4. Turn off the burner and place the skillet into the oven for 2 minutes, until the marshmallows have puffed up and are lightly browned.

5. Remove from the oven and immediately serve. Use your graham crackers as single-dip spoons, and shovel a heap of the gooey chocolate and marshmallow perfection into your mouth.

Equipment

USA Skillet by American Skillet Company

Ingredients

1 Tbsp. of butter

dark chocolate chips

jumbo marshmallows

graham crackers for serving

Pineapple Upside-Down Cake
Butter Pat Industries

Pineapple upside-down cakes are often made in cast-iron Bundt pans, but this recipe utilizes a simple cast-iron skillet. Butter Pat Industries says this delicious recipe is credited to Carlos Greenwood's mother. Thank you, Mrs. Greenwood!

Directions

1. Preheat oven to 350°F.

2. Drain pineapple and save the syrup.

3. Heat the skillet on the stove and melt enough butter to cover the bottom of the pan (approximately 3 tablespoons). Add brown sugar to cover bottom of pan. Add drained pineapple slices to cover the brown sugar. Place Maraschino cherries in pineapple slices.

4. In a bowl, sift together flour, baking powder, and salt.

5. In a separate bowl, mix together the remaining ingredients plus 5 tablespoons of the reserved syrup. Beat this into the flour mixture and then pour over the pineapple.

6. Bake approximately 35–40 minutes. Allow to cool. Place a large plate on top of the skillet then flip the cake onto the plate for serving.

Equipment

10" Butter Pat cast-iron skillet (known as the Heather)

Ingredients

2 8-oz. cans sliced or crushed pineapple in heavy syrup

8 Tbsp. (1 stick) butter, divided

⅓ cup brown sugar

Maraschino cherries (optional)

1 cup cake flour (not self-rising cake flour)

1½ tsp. baking powder

½ tsp. salt

¼ cup butter-flavored shortening

¾ cup granulated sugar

½ cup half-and-half

1 large egg, room temperature

2 tsp. vanilla extract

Notes

COLLECTING VINTAGE CAST IRON

1. Lodge Cast Iron, "The History of Lodge."
2. Cast Iron Collector, "Factory Automation."
3. Cast Iron Collector, "Gate Marks."
4. Cast Iron Collector, "Ghosts in the Machine."
5. Gear Patrol, "There's a Very Tricky Ring on the Bottom of Your Cast-Iron Skillet. What's It for?"
6. Journal of Antiques & Collectibles, "There's History in Your Frying Pan."
7. Bhamwiki, "Birmingham Stove & Range Company."
8. Made in Chicago Museum, "Chicago Hardware Foundry Co. & Harper Supply Co., est. 1897."
9. Made in Chicago Museum, "Chicago Hardware Foundry Co. & Harper Supply Co., est. 1897."
10. Boonie Hicks, "The Favorite, Vintage Cast Iron by the Columbus Hollow Ware Co."
11. Worthpoint, "Favorite Stove and Range Co."
12. Boonie Hicks, "Favorite Piqua Ware: Favorite Stove and Range Co. Vintage Cast Iron."
13. Boonie Hicks, "Griswold Cast Iron Skillet. Simple Identification Guide Using Logos."
14. Vollrath, "Vollrath Heritage."
15. Smith and Wafford, *The Book of Griswold & Wagner.*
16. Smith and Wafford, *The Book of Griswold & Wagner.*
17. Smith and Wafford, *The Book of Griswold & Wagner.*

COLLECTING MODERN CAST IRON

1. Thomas, "What Is CNC Machining?"
2. Gear Patrol, "There's a Very Tricky Ring on the Bottom of Your Cast-Iron Skillet. What's It For?"

RESTORING CAST IRON

1. Metal Detecting World, "Rust Removal by Electrolysis."
2. The Editors at Test Kitchen, "The Science of Seasoning," *Cook It in Cast Iron.*

Bibliography

Anderson, James P. *A Cast Iron Journey*. Texas, 2019.

Bhamwiki. "Birmingham Stove & Range Company." Modified May 12, 2017. https://www.bhamwiki.com/w/Birmingham_Stove_%26_Range_Company.

Boonie Hicks. "Favorite Piqua Ware: Favorite Stove and Range Co. Vintage Cast Iron." Accessed February 19, 2022. https://www.booniehicks.com/favorite-piqua-ware/.

Boonie Hicks. "The Favorite, Vintage Cast Iron by the Columbus Hollow Ware Co." Accessed February 19, 2022. https://www.booniehicks.com/the-favorite-columbus-hollow-ware-co.

Boonie Hicks. "Griswold Cast Iron Skillet. Simple Identification Guide Using Logos." Accessed February 19, 2022. https://www.booniehicks.com/griswold-cast-iron-skillet.

Cast Iron Collector. "Factory Automation." Accessed February 19, 2022. http://www.castironcollector.com/automation.php.

Cast Iron Collector. "Gate Marks." Modified June 8, 2015. https://www.castironcollector.com/forum/showthread.php?t=2213.

Cast Iron Collector. "Ghosts in the Machine." Accessed February 19, 2022. http://www.castironcollector.com/ghosts.php.

The Editors at Test Kitchen. *Cook It in Cast Iron: Kitchen-Tested Recipes for the One Pan That Does It All*. Edited by America's Test Kitchen. Brookline, MA: Cook's Country, 2016.

Gear Patrol. "There's a Very Tricky Ring on the Bottom of Your Cast-Iron Skillet. What's It for?" Modified February 12, 2022. https://www.gearpatrol.com/home/a724173/ring-bottom-cast-iron-skillet.

Jones, Ashley. *Modern Cast Iron: The Complete Guide to Selecting, Seasoning, Cooking, and More*. Bloomington, IN: Red Lightning Books, 2020.

Journal of Antiques & Collectibles. "Hearth to Hearth: There's History in Your Frying Pan." Last modified January 2001. https://journalofantiques.com/misc/hearth-to-hearth-theres-history-in-your-frying-pan.

Lodge Cast Iron. "The History of Lodge." Accessed February 19, 2022. https://www.lodgecastiron.com/about-lodge/history.

Made in Chicago Museum. "Chicago Hardware Foundry Co. & Harper Supply Co., Est. 1897." Accessed February 19, 2022. https://www.madeinchicagomuseum.com/single-post/chicago-hardware-foundry-co.

Metal Detecting World. "Rust Removal by Electrolysis—A Detailed Illustrated Tutorial, Page 2." Accessed February 19, 2022. https://www.metaldetectingworld.com/electrolysis_rust_removal_p2.shtml.

Smith, David G., and Chuck Wafford. *The Book of Griswold & Wagner: Favorite, Wapak, Sidney Hollow Ware*. Revised and Expanded Fifth Edition. Atglen, PA: Schiffer Publishing, Ltd., 2013.

Smith, David G., and Chuck Wafford. *The Book of Wagner & Griswold: Martin, Lodge, Vollrath, Excelsior*. Atglen, PA: Schiffer Publishing, Ltd., 2001.

Thomas. "What Is CNC Machining? Definition, Processes, Components, & More." Accessed February 19, 2022. https://www.thomasnet.com/articles/custom-manufacturing-fabricating/understanding-cnc-machining.

Vollrath. "Vollrath Heritage." Accessed February 19, 2022. https://vollrathcompany.com/about-us/heritage.

Worthpoint. "Favorite Stove and Range Co." Accessed February 19, 2022. https://www.worthpoint.com/dictionary/p/metals/manufacturers-american/favorite-stove-and-range-co.

Index

Ashley L. Jones is a firm believer that we should share what we know with others . . . and she knows a lot about cast-iron cookware. It all started in 2010 when her future in-laws began filling her head with family stories of cooking hoe cakes on a cast-iron stove and baking biscuits in Dutch ovens over hot coals. When they gave her a two-quart Dutch oven of her own, she felt like part of the family.

Later, Ashley received a couple of refurbished skillets from her new in-laws. After using them for a while, though, food started to stick to the pans. That's when Ashley began to explore different methods of cleaning and seasoning cast iron, ultimately developing her own system.

Along the way, Ashley heard that some cookware leaches chemicals into food. After investigating the issue, she learned that cast iron is one of the healthiest cookware options available today. Armed with yet another reason to use her black pans, she started cooking in them on a daily basis.

Over the years, Ashley has continued to research the history, use, and care of cast iron and share what she's learned with others. She loves to experiment with new recipes and challenges herself to re-create old favorites in one of her skillets or Dutch ovens.

To Ashley, cast iron is more than just cookware. It's a tangible connection to a simpler time and a slower pace. Her goal with this book and her first release, *Modern Cast Iron*, is to encourage readers to grab hold of that tradition, to cook wholesome foods and gather with family around the table.